The Michigan Historical Reprint Series

Reprints from the collection of the University of Michigan University Library

The Scholarly Publishing Office
the University of Michigan
University Library

THE NEW YORK ELECTRICAL HAND BOOK

THE NEW YORK
ELECTRICAL HANDBOOK

THE NEW YORK ELECTRICAL HANDBOOK

Being a Guide for Visitors from Abroad Attending the International Electrical Congress, St. Louis, Mo. September, 1904

New York
Published under the auspices of
The American Institute of Electrical Engineers
1904

THE MASON PRESS
Syracuse, New York

CONTENTS

INTRODUCTION

The New Bridge Across the East River

Introduction

NEW YORK CITY, the metropolis of the United States and the largest and wealthiest city in the Western Hemisphere, is finely situated on the islands and mainland at the mouth of the Hudson River. The population of the city in 1903 was estimated by the United States Census Bureau at 3,716,139. In point of numbers, as well as from a commercial and financial standpoint, the city is second to London alone among the great cities of the world. Although owing to the configuration of State boundaries certain neighboring cities, such as Jersey City, Newark and Paterson, lie in the State of New Jersey and can not be reckoned as a part of the city in a political sense, they should, nevertheless, be considered in the Metropolitan area, and with this addition the total population is increased to about 4,500,000.

The original settlement was made in the southern end of Manhattan Island in 1624 by the Dutch, and was known as New Amsterdam. After its capture by the English in 1664, the city, which had grown as far up the island as Wall street, was renamed New York, in honor of James, Duke of York, to whom the province was granted by his brother, Charles II. One hundred years later, just before the outbreak of the Revolution, the city extended to where the City Hall now stands.

In 1858 the Fifth Avenue Hotel stood at the northern end of the city, but the opening of the elevated railways between 1877 and 1880 gave a great impetus to the northward growth of the city, until to-day the whole of Manhattan Island, from end to end, is densely populated, with the exception of the spaces reserved for public parks.

The present city of New York consists of five boroughs, namely, Manhattan, Bronx, Brooklyn, Queens and

Richmond. Its length from north to south is thirty-three miles; its width from east to west, eighteen miles, and its total area 307 square miles.

The Borough of Manhattan is conterminous with the island of that name. It is long and narrow, and lies nearly north and south in the axis of its greatest length, the southern and oldest settled portion of the island facing the traveler as he approaches through New York harbor from the Atlantic Ocean. The extreme length of the island is thirteen and one-half miles and its breadth varies from half a mile to two miles and a half, the average being about a mile and three-quarters. Its total area is twenty-one square miles.

In 1874, an area equal in size to Manhattan Island was taken from Westchester county on the mainland, west of the Bronx River, and added to the city, and in 1895 this area was further extended to include all the territory between the Hudson River and Long Island Sound, south of the cities of Yonkers and Mount Vernon.

On January 1st, 1898, the city of Brooklyn, Long Island City and part of Queens County, and Staten Island, the boundaries of which are conterminous with the county of Richmond, were incorporated in New York City. The area and population of the city and of its five boroughs at the official United States census of 1900 were as follows:

	Area sq. miles.	Population.
New York City	307	3,437,202
Manhattan Borough	21	1,850,093
Bronx "	39	200,507
Brooklyn "	66	1,166,582
Queens "	124	152,999
Richmond "	57	67,021

The growth of the population of New York City since 1900 is estimated by the United States Census at not less than 100,000 a year, a considerable proportion of which is due to foreign immigration. It has been estimated that the population of the metropolitan area of New York contains 900,000 inhabitants of German birth or parentage; 850,000 inhabitants of Irish birth or parentage; 200,-

Standard Oil Building

000 inhabitants of English, Scotch or Canadian parentage; 100,000 inhabitants of Italian birth or parentage, and 100,000 inhabitants of Russian birth or parentage, mostly Hebrews. There are also about 10,000 Chinese, and smaller colonies of French, Spaniards, Greeks, Syrians, Armenians and Japanese.

The government of New York City is of the representative American type. The Mayor is the chief executive, and is elected by popular suffrage for a term of two years at a salary of $15,000 per year. He appoints, and may remove at will, all executive heads of departments. The principal offices which the Mayor fills by appointment are as follows: Police Commissioner, Street Cleaning Commissioner, Commissioner of Bridges, Commissioner of Public Charities, Commissioner of Correction, Fire Commissioner, Health Commissioner, Building Commissioner, Tenement House Commissioner, three Park Commissioners, five Commissioners of Taxes and Assessments, City Chamberlain, Corporation Counsel, and the Commissioner of Water, Lighting, Gas and Supplies. Each of these commissioners is head of the department which the name of his office indicates, and the responsibility for its management rests absolutely with him. Besides the Mayor, the Controller, who has charge of the finances, and the President of the Board of Aldermen, who takes the place of the Mayor in his absence, are elected in the same manner as the Mayor and for a similar term. The Controller receives a salary of $15,000 per year, and the salaries of the commissioners average $7,500 each.

The legislative department of the city government is vested in a Board of Aldermen, which consists of a single chamber of seventy-three members, who are elected from specified districts. This board has restricted powers of legislation and takes no active part in the executive administration of the city government. Each of the five boroughs elects a President by popular vote, and each President in turn selects a Commissioner of Public Works for his district.

The most important bureau of the city government is

the Board of Estimate and Apportionment, which passes on all expenditures made by the city, and prepares the annual budget. This board consists of the Mayor, Controller, and President of the Board of Aldermen, each with three votes, the Presidents of the Boroughs of Manhattan and Brooklyn with two votes each, and the other three Borough Presidents with one vote each.

Since the boundaries of New York were extended to include Brooklyn, part of Queens County and Staten Island, the annual appropriations have nearly reached $100,000,000 each year. It is characteristic of an American city that the largest single item in the appropriation should be devoted to education, the amount for 1903 being $20,063,017.77; in addition to which there was also appropriated $298,362 for the College of the City of New York, where male teachers are trained, and $220,000 for the Normal College of the State of New York, where female teachers are trained. Next to education the largest amount spent was for interest on the city debt, amounting to $13,276,709.64, together with the sum of $10,417,-359.17 paid into the sinking fund for the ultimate redemption of the debt. The Police Department, which has about 7,000 members, absorbed $11,566,680.42, while the Fire Department, with 3,000 members, and the Street Cleaning Department required a little over five million dollars each.

The total assessed valuation of the entire city is $4,764,205,484, of which the Borough of Manhattan is assessed at $3,507,083,911. The city debt, deducting sums paid into the sinking fund, amounts to a little over two hundred million dollars. It would appear, therefore, that the assessed valuation of New York is almost equal to that of London, while the debt is about the same. There are about 1,300 miles of streets, of which 1,100 miles are paved, and 1,100 miles of sewers. Owing to the scouring effects of the tides in the Hudson River, New York has not had the same difficulties to face as many other large cities in getting its sewage out to sea.

The traveler from Europe approaching New York sails almost due west until he passes Sandy Hook and

Scotland lightships. Sandy Hook itself is a low sand spit extending out from New Jersey, on which is situated

J. P. Morgan & Co.

Fort Hancock, defended by large, modern, high-power guns mounted on disappearing carriages. After crossing the bar, which extends from Sandy Hook to Long Isl-and and which has two channels, the vessel turns sharply almost due north and enters the lower bay. On the left hand side is then seen the wooded heights and villas of Staten Island, while on the right hand side Coney Island and Rockaway, two seaside resorts in the Borough of Brooklyn, come into view. The vessel now passes through the Narrows, about a mile and a half wide, formed by Staten Island on the left and Long Island on the right, and guarded by Fort Wadsworth on the Staten Island side and Fort Hamilton on the Long Island side. Just within the Narrows is the quarantine station, at which all ships stop for examination by the health officers. After passing quarantine the vessel enters the upper bay and the traveler gets his first glimpse of Manhattan Island, the lower part of which presents a sight at present unequaled in the world, owing to the number and diversity of the tall buildings which cluster around its southern end. The first object which attracts the attention of the traveler in the upper bay is the Statue of Liberty, a colossal statue 151 feet high and standing upon a pedestal 155 feet in height, designed by Bartholdi. On the right, almost opposite to the Statue of Liberty, may be seen the beautiful Greenwood Cemetery, situated in the Borough of Brooklyn. Further along to the right may be seen the light and graceful structure of the Brooklyn Bridge, spanning the East River, which is really an arm of the sea connecting the upper bay with Long Island Sound. The vessel passes up the left hand side of Manhattan Island through the North River, as the Hudson River is here locally known, and on the Jersey shore may be seen the numerous railway stations of the great trunk lines leading to the west and south, including first the Central of New Jersey, then the Pennsylvania, the Lackawanna, and the Erie railways. At the southern end of Manhattan Island may be seen a patch of green, twenty-one acres in extent, known as Battery Park. Within the limits of this park stands Castle Garden, originally a fort, then an amusement hall in which Jenny Lind sang in 1850, after that a depot for receiving immigrants, and now a public aquarium. On the right may be seen the Produce Exchange, a large building in red brick and terra cotta, in the Italian Renaissance style, with a square tower two hundred and twenty-five feet high. To the left of the Produce Exchange, and somewhat beneath it, will stand the new Custom House, now in process of erection, on the south side of Bowling Green. On the left hand side is the Washington Building, another red

The New Hall of Records

brick structure, erected by the late Cyrus W. Field, which is the beginning of Broadway.

The vessel now passes to her pier either on the New York or New Jersey side of the North River and after a somewhat troublesome examination by the custom house officers the traveler makes his way to his hotel.

The visitor to New York will find that the best way to obtain a general view of the city is to start at the lower end of Broadway and walk as far up as the City Hall, barely a mile distant, take a street car from that point to Madison Square and 23rd street, and then take a stage up Fifth avenue to Central Park.

Beginning at No. 1 Broadway, on the left hand side, is the Washington Building, already mentioned, and next to it is the large Bowling Green Building. On the other side of the street, opposite to these two buildings, is the Welles Building and the Standard Oil Building. At the southern right hand corner of Exchange place is the Exchange Court Building, with four large bronze statues: of Stuyvesant, the last Dutch governor of New Amsterdam; of DeWitt Clinton, the originator of the Erie Canal; of Henry Hudson, the discoverer of the Hudson River, and of James Wolfe, one of the two heroes of the Heights of Abraham. On the north right hand corner of Exchange place is the Consolidated Stock and Petroleum Exchange, and a few doors above this at Nos. 64 to 68 is the building of the Manhattan Life Insurance Company, the tower of which is 350 feet in height. On the left hand side and at the corner of Rector street, which separates it from Trinity churchyard, is the Empire Building, twenty stories high.

Trinity Church, which stands opposite the head of Wall street, is a fine Gothic edifice of brown stone with a spire 285 feet high. The present church was completed in 1846, but it stands on the site of two earlier churches, the first of which was erected in 1696. The handsome bronze doors of this church were given as a memorial of the late John Jacob Astor by his descendants, and the altar and reredos in the church itself were built as a memorial of William B. Astor, son of John Jacob Astor the first. The churchyard contains the graves of Alexander Hamilton, the celebrated American statesman who, in the words of Talleyrand, "divined Europe;" Robert Fulton of steamboat fame; Captain Lawrence, who was killed in the Chesapeake-Shannon fight in 1813; General Phil Kearney, and William Bradford, who died in 1752, printer of the first New York newspaper.

The visitor now turns down Wall street and sees on the left the United States Sub-Treasury, a marble building with a Doric portico. This building occupies the site of the old Federal Hall, in which the first United States Congress was held and in which Washington was inaugurated President. A large bronze statue of Washington by J. Q. A. Ward, erected in 1883, marks the site where Washington took the oath as first President of the United States in 1789. Opposite the Sub-Treasury and running south parallel to Broadway is Broad street, on which is the New York Stock Exchange, a handsome marble structure; at the right hand corner of Wall and Broad streets is situated a white marble building known as the Drexel Building, on the ground floor of which J. P. Morgan and Company have their offices. Further down Broad street, on the right hand side, we first pass, at No. 20, the Commercial Cable Building, and, at No. 44, the Edison Building, which is the New York headquarters of the General Electric Company. Resuming our walk through Wall street, we see the United States Assay Office, where crude bullion is refined, and further along to the right we see the huge Ionic colonnade of dark granite of the present Custom House, but which will be occupied by the City Bank as soon as the new Custom House on Bowling Green is completed.

Resuming our walk up Broadway, we pass the twenty-three-story building of the American Surety Company, 316 feet high, and at the corner of Pine street and on the same side, between Pine and Cedar streets, the huge building of the Equitable Life Assurance Society with 1,500 tenants. It is in this building that the New York offices of the Westinghouse Company are situated. The offices and library of the American Institute of Electrical Engineers are situated at No. 95 Liberty street, which crosses Broadway one block above the Equitable Building. At No. 195 Broadway, corner of Dey street, on the left hand side of the street, is the large building of the Western Union Telegraph Company, and down Dey street, at No. 15, is the chief office of the New York Telephone Company. Higher up, on the west side of Broadway, is St. Paul's

Church, one of the earliest chapels of the Trinity Church corporation, which was erected in 1765 and is the oldest church edifice in New York City. Here Washington worshipped when he was President, during his first term, while the Federal Government was situated in New York City, and his pew is still shown to visitors. Opposite St. Paul's Church is the St. Paul Building with its twenty-six stories, 308 feet high, and above this and to the right is the huge Park Row Building, 390 feet high, containing 950 offices and twenty-nine stories.

At the apex of a triangle formed by Park Row and Broadway stands the Post Office, architecturally a very unsatisfactory building, completed in 1876. Opposite to the Post Office, on the left hand side of Broadway, stands the Astor House, erected in 1834 and long the most famous hostelry in New York, although now somewhat too far down town. This building covers the site of the first house occupied by John Jacob Astor, to whose descendants the hotel still belongs.

Above the Post Office is an open space laid out in a small park, which is all that remains of the commons of the city. Here in the early Revolutionary days the patriots drilled, and it was here that George Washington read to the assembled troops the Declaration of Independence a few days after it had been signed in Philadelphia. The City Hall, perhaps the most satisfactory building in an architectural sense in New York, occupies the centre of this park. It was finished in 1812 and its architect was John McComb. The north side of the building was faced with freestone, as no one then dreamed that the city would ever extend beyond this point. North of the City Hall stands the Court House, a white marble building with Corinthian columns, noted as having been the chief means through which the stupendous Tweed frauds were realized. Its cost on account of these frauds has been variously estimated at from $12,000,000 to $17,-000,000. On the Park Row side of City Hall Park four newspaper buildings stand out conspicuously. The first of these, going from south to north, is the building lately

The Flatiron Building

occupied by the New York *Times,* the architect of which was Mr. George B. Post, and which is probably the most satisfactory architectural treatment of a tall building to be seen in the city. The next is the New York *Tribune* Building, one of the earliest attempts at a tall building in the country, with a clock tower 285 feet high. The New York *Sun* Building, of no architectural pretensions, stands on the same block as the *Tribune* Building; it is noted as having been the former home of the Tammany Society. Further on is the Pulitzer Building, the office of the New York *World,* with a dome 310 feet high, from which a splendid view of New York may be obtained. Beyond the Pulitzer Building is the entrance to the East River Bridge, usually called the Brooklyn Bridge.

The northern boundary of City Hall Park is made by Chambers street, at the eastern end of which is the new Hall of Records, a handsome building in the Corinthian style. On the northeast corner of Chambers street and Broadway is the Stewart Building, erected by the late A. T. Stewart for his wholesale dry goods business, but now given up to offices, many of which are rented by the city.

On the Broadway side of City Hall Park stands the fine building of the Postal Telegraph Company and the offices of the Home Life Insurance Company. A few blocks north of Warren street is Duane street, leading to the offices and one of the power stations of the New York Edison Company. The building runs through to Pearl street and replaces an earlier building further down Pearl street, where Mr. Edison, in 1881, erected one of the earliest stations for the distribution of low tension current for incandescent electric lighting. At the corner of Broadway and Leonard street are the huge offices of the New York Life Insurance Company. Higher up on Broadway, at No. 621, at the intersection of Houston street, two miles from the Battery and one mile from City Hall, is the Cable Building, in which the offices of the New York City Railway Company, formerly the Metropolitan Street Railway Company, are situated. This company controls all the street car lines in the Boroughs of Manhattan and the Bronx.

Houston street was formerly known as North street, and early in the last century marked the extreme limit to which the straggling buildings of the upper part of the city extended. A commission was appointed in 1807 to lay out the upper streets of the city and Houston street was taken as the starting point. It was decided to lay out streets designated by numbers instead of names, running east and west and making twenty to the mile, and to intersect these streets by numbered avenues, running north and south, giving seven to the mile. A slight protuberance on the eastern side of Manhattan Island below Fourteenth street was taken care of by short avenues designated A, B, C and D. Later on two additional avenues were added, one to bisect the territory between Fifth and Fourth avenues, called Madison, and one to bisect the territory between Fourth and Third avenues, called Lexington avenue. The streets are called "East" or "West" according as they lie east or west of Fifth avenue. North of Houston street, fronting on the Hudson River, formerly stood the country house and grounds of Admiral Sir Peter Warren, who married a daughter of James De Lancey, the largest owner of real estate in the New York City of Colonial times, as the Astor family has been in post-Colonial days. Sir Peter called his residence "Greenwich" because it was up the river, and after the Revolution his estate was cut up into streets and building lots which did not fit in with the chessboard scheme of the Street Commission of 1807. On this account it is not until we reach Fourteenth street that the streets run regularly clear across Manhattan Island. Above Fourteenth street all the streets on Manhattan Island preserve their regularity as already explained.

From Bowling Green to Tenth street, a distance of approximately two and a half miles, Broadway runs as straight as an arrow, its direction being a little east of north. At Ninth street and Broadway is the beautiful retail store erected by A. T. Stewart a few years before his death in 1876. This is one of the few examples of a building completely encased in cast iron now remaining. It is now occupied by John Wanamaker.

Beyond Ninth street, Broadway inclines to the left, running almost exactly due north, and at the bend stands Grace Church, a beautiful pile of buildings in white limestone with a marble spire, designed by the late James Renwick. At Fourteenth street Broadway reaches Union Square, a handsome pleasure ground with several statues, including an equestrian statue of Washington by H. K. Browne and a bronze statue of Lafayette by Bartholdi.

Fourteenth street is usually taken as marking the boundary line between down-town and up-town. The southern end of Manhattan Island as far up as the City Hall is given up to finance in its various forms—banking, the purchase and sale of stocks and bonds, insurance, the headquarters of the great industrial corporations and of the many great railway companies. Between the City Hall and Fourth street, Broadway itself and the area lying immediately west of it is given up to the wholesale dry goods trade. East of Broadway, however, between the City Hall and Fourteenth street, lie some of the most densely populated districts in the world, partly packed away in huge tenement houses and partly in houses which were once middle-class residences, but which have descended through the various stages of "genteel" lodgings, of cheap boarding houses, of houses where light housekeeping is permitted, to the final stage of houses where eight or ten newly arrived immigrants from Southern and Eastern Europe occupy a room jointly. The East Side above the City Hall toward the East River contains probably the largest Hebrew population in any city in the world. The lower end of Mott street is given up to Chinese, and the lower end of Mulberry street to Italians. The great artery for this heterogeneous population is the Bowery, a broad thoroughfare, which runs roughly parallel to Broadway and at some distance to the east of it, from Chatham Square to the Cooper Institute at Eighth street. The changes which have passed over the Bowery in seventy-five years are typical of what is happening elsewhere in many other parts of New York. Before the

Irish famine in 1847 the lower end of the Bowery and the side streets east of it were inhabited chiefly by Quakers, while a Methodist Church was the centre of social life at its upper end. Between 1850 and 1880 the Irish immigrants steadily drove out the original residents. These were the days of the volunteer firemen and of "the Bowery boy," which made Tammany Hall almost a purely Irish organization, and which gave the Bowery throughout the United States a somewhat unenviable reputation for fun and frolic. In 1880 the total number of alien immigrants into the United States sprung suddenly from 177,826 the previous year to 457,257 for that year, and since then the average number of immigrants has been about a half million a year. Nor was this the only change. Previous to 1880 the immigrants into this country were chiefly drawn from the northern races of Europe, Germany and Ireland being conspicuous among these. Since 1880 the great bulk of immigrants have been from Italy, Austria-Hungary and Russia. These immigrants in turn have driven out of the lower East Side the previous Irish and German residents, so that this part of New York is a huge foreign city in itself, speaking several tongues other than English. Fortunately, the splendid public school system of New York shows its wonderful assimilating effect on the children of these immigrants, who quickly learn English and in many cases completely forget the tongue of their parents. These changes on the lower East Side have had their effect on the Bowery, which, although still to a large extent a region of small stores, cheap playhouses, museums and saloons, is now virtually in a transition stage, and the probabilities are that before many years the great width of this street will be taken advantage of by manufacturers and wholesale dealers.

Returning now to Union Square, we find that Broadway makes a sharp turn to the left, passing around the left hand side of the Square. At the corner of Sixteenth street and Union Square is situated at present the splendid store of Tiffany & Company, the jewelers *par excel-*

lence of the Western Hemisphere. This store, following the trend of many other high-class stores in New York City, is about to move up-town to a site on Fifth avenue. Here are also the New York offices of the Raymond & Whitcomb Company. Between Union Square and Madison Square, which lies to the right of Broadway, beginning at Twenty-third street, are situated many of the finest dry goods and furniture stores in the city. Twenty-third street itself is a wide thoroughfare with street cars running from river to river, and the block between Fifth and Sixth avenues contains many of the best retail stores in the city.

Where Fifth avenue and Broadway intersect at Twenty-third street, and lying between that street and Twenty-second street, stands the Flatiron Building, one of the most remarkable buildings in the world. It measures 171 feet on the Fifth avenue side by 86 feet on the Twenty-second street side, these two sides forming the perpendicular and base of a right angle triangle, the hypothenuse of the triangle being completed by the Broadway side. The building is twenty-one stories, or about 285 feet, above the curb. On account of its exposure and on account of its great height and extreme narrowness at the north end, it was necessary to make suitable provision to withstand the great wind stress, and its steel construction is thoroughly braced and stiffened at each story by knee braces.

On the northwest corner of Fifth avenue and Twenty-third street is situated the Fifth Avenue Hotel, which succeeded the Astor House as the leading Hotel in New York and easily held its preëminence until a few years ago, when several larger and finer hotels were erected further up Fifth avenue. Twenty-third street also marks the junction point where Broadway, still inclining to the left, crosses Fifth avenue. At Thirty-fourth street Broadway crosses Sixth avenue; at Forty-second street it crosses Seventh avenue; and at Fifty-ninth street, the southwest corner of Central Park, Eighth avenue.

Between the Battery and Forty-seventh street, a distance of almost four and a half miles, Broadway is one of the great thoroughfares of the world, lined with splendid stores and offices and its sidewalks filled with bustling crowds of people. As already described, at its lower end are the office buildings of the large financial and corporate interests of New York; in its middle portion, be-

The Ansonia, Broadway and 73rd Street

tween City Hall and Fourteenth street, it is chiefly given up to wholesale dress goods and clothing establishments. Between Fourteenth and Thirty-third streets are many fine retail stores and hotels and some of the principal theatres. Between Thirty-third and Forty-seventh streets it is almost entirely given up to hotels and theatres.

Above Fifty-ninth street, Broadway widens into a magnificent boulevard which continues up to the end of Manhattan Island.

Fifth avenue, which has long been the fashionable residence street of New York, begins at Washington Square and extends to the Harlem River, which it reaches at One Hundred and Forty-third street. Between Washington Square and Thirteenth street there are still some fine old-fashioned residences, but from Thirteenth street to Forty-second street it is now almost entirely given up to expensive stores, hotels and clubs. Between Twenty-third and Twenty-sixth streets it skirts Madison Square, where a fine statue of Admiral Farragut by Saint-Gaudens may be seen. At the corner of Twenty-ninth street is the Calumet Club on the right hand side, and on the left hand side at the corner of Thirtieth street is the Holland House, one of New York's leading hotels. The Knickerbocker Club stands at the corner of Thirty-second street to the right, and the entire block on the left between Thirty-third and Thirty-fourth streets is occupied by the Waldorf-Astoria Hotel, one of the largest and finest hotels in the world. On the northwest corner of Thirty-fourth street stood for a quarter of a century the former costly residence of A. T. Stewart, afterwards turned into a club and now torn down and replaced by several buildings, prominent among which is that of the Knickerbocker Trust Company.

Between Thirty-fifth and Thirty-sixth streets on the west side is situated at present the Engineers' Club, soon to be removed to a much finer site on Fortieth street opposite the new Public Library.

The new store of Tiffany and Company is situated on the southeast corner of Thirty-seventh street and Fifth avenue.

The Union League Club stands on the northeast corner of Thirty-ninth street and is the principal Republican club in the United States and perhaps the leading political club in the country.

Between Fortieth and Forty-second streets, on the west side of Fifth avenue and extending with its grounds to Sixth avenue, will stand the New York Public Library (Astor, Lenox and Tilden foundations), now in course of erection.

At the southeast corner of Forty-fifth street stands Sherry's, and on the northeast corner of the same street Delmonico's, two of the famous restaurants of the city.

At Forty-ninth street on the east side the Buckingham Hotel begins, extending through to Fiftieth street. Between Fiftieth and Fifty-first streets stands St. Patrick's Cathedral, a large white marble edifice designed by James Renwick. On the block above the Cathedral stands the Union Club, one of the oldest and still the most fashionable club in the city.

Between Fifty-first and Fifty-second streets on the west side are two Vanderbilt residences of brown stone, while on the corner of the next street above is the residence of Mr. William K. Vanderbilt, a handsome house in the French chateau style.

The University Club is situated on the northwest corner of Fifty-fourth street.

Between Fifty-seventh and Fifty-eighth streets on the west side is the splendid residence of Mrs. Vanderbilt, widow of the late Cornelius Vanderbilt.

The large square in front of the entrance to Central Park is occupied by three hotels, namely, the Plaza, the Netherland, and the Savoy.

Central Park extends from Fifty-ninth street to One Hundred and Tenth street and from Fifth to Eighth avenue, and these boundaries have an exclusive area of 840 acres. Central Park presents a beautiful variety of wood, water and lawn and very skillful use has been made of its natural features to make them afford the greatest extent of roads for driving, for horseback exercise and of paths for pedestrians.

Fifth avenue where it extends along the east side of Central Park has not been encroached upon by business houses and many fine residences stand in this section of the Avenue, including that of Mr. J. J. Astor at Sixty-fifth street and Mr. Andrew Carnegie on the block between Ninetieth and Ninety-first streets.

The Metropolitan Museum of Art, situated just within the eastern wall of Central Park between Eighty-second and Eighty-third streets, contains many fine pictures and has by far the best art collection in the United States.

The beautiful Riverside Drive is a park which extends from Seventy-second street to One Hundred and Twenty-seventh street. Many handsome residences are built along the eastern side of this park, while its western side runs down to the shores of the Hudson River, of which it commands a magnificent view.

THE BRIDGES ACROSS THE EAST RIVER

An enormous population has grown up on the Long Island shore of the East River which has made it necessary to build two bridges and begin work on two more, so that the tedious travel by ferries across the river can be avoided. The streets on the east side of the river in Brooklyn, where the larger part of this population is centered, converge at two places, and the two bridges already completed were constructed at these points. The third bridge is considerably farther north, at a place where an island renders construction relatively easy, and when it is finished it will open up a district on Long Island which will be rapidly taken up for the homes of men doing business uptown in Manhattan. The fourth bridge will be erected close to the first, mainly in order to relieve it of the traffic which overcrowds it at present.

The first structure to be built was the Brooklyn Bridge, erected in 1870-83. It was designed by John A. Roebling, who was succeeded successively by Washington A. Roebling and C. C. Martin as chief engineer. It is a suspension bridge with a main span of 1,595 feet 6 inches and a land span of 930 feet at each end. Four cables, each 15¾ inches in diameter, are used to carry the roadways and tracks, and six rudimentary stiffening trusses are provided. The extreme width of the bridge is 85 feet, and it carries a footway of 15½ feet width, two tracks for the cars of the elevated railways and the railway system of the bridge itself, and two 18-foot roadways, each of which is considerably obstructed by a track for trolley cars. The arrangement of roadways differs materially from that which was contemplated when the general designs for the bridge were prepared, and numerous projects for increasing the strength of the structure and consequently its capacity, have been suggested. The special engineering feature of the bridge is the use of masonry towers. These were built on pneumatic caissons which were the largest of their kind when they were sunk. One tower is 140 by 59 feet at the water line and the other is 140 by 56 feet. Each is 272 feet above high water. The Brooklyn tower contains 38,214 cubic yards of masonry and the New York tower 46,945 cubic yards.

The bridge extends with its approaches from Park Row, Manhattan, to Sands street, Brooklyn, a total distance of 6,016 feet, and if the extreme points of the terminal structures at the ends are used in the measurement the length is 7,580 feet. This bridge crosses the river near what was the most important ferry, before its construction, between the two cities.

The Williamsburg Bridge, also a suspension structure, connects the foot of the main street of the old city of Williamsburg with Delancy street, Manhattan, which is close to the centre of a district employing a large part of the population having its home in Williamsburg and formerly crossing the river by three ferries. This bridge has a main span of 1,600 feet and a land span at each end of 596½ feet. These land spans are not suspended from the cables like those of the Brooklyn Bridge, but are independent truss structures.

The towers, rising about 332 feet above the high water level, are steel structures weighing about 3,048 tons each. Instead of one large caisson and a single base for each tower, as were used for the Brooklyn Bridge, two caissons and two masonry piers were employed for each tower of the Williamsburg Bridge, and one of these caissons had to be sunk to a depth of 107½ feet, thirty feet deeper than at the older bridge.

The towers carry four cables, each 18⅝ inches in

diameter. Two stiffening trusses 40 feet deep form one of the most noticeable features of the structure, and, when viewed from a distance, distinguish it more than any-

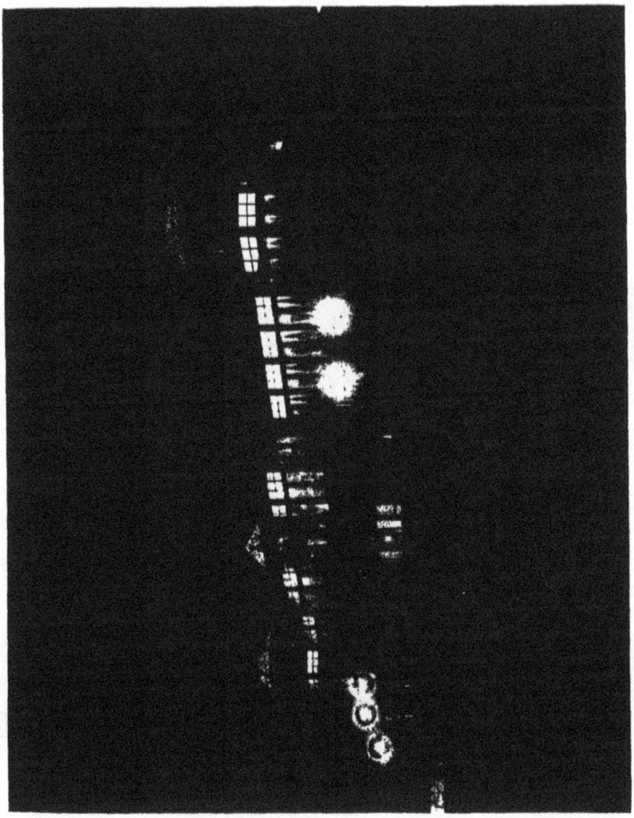

"Casino," Central Park

thing else from the Brooklyn Bridge. The topographical conditions at each end of the bridge rendered it advisable to keep all the heavy traffic on one level, and for this reason the bridge has a width of 118 feet. A light overhead deck carries the footwalks and bicycle paths, while the main floor has two tracks for elevated railway cars, four tracks for trolley cars and two twenty-foot roadways; the length of these last from the terminal at one approach to that of the other is 7,264 feet. The bridge has about three times the traffic capacity of the Brooklyn Bridge.

The bridge was designed by L. L. Buck and built under his supervision, O. F. Nichols being his principal assistant engineer during all the important stages of the work. It was started in 1895 and was practically finished in 1903.

The third bridge differs from the others in being a cantilever structure, a type of construction rendered possible by the natural advantages for tower foundations of Blackwell's Island, over which it passes and from which it derives its name. Beginning on the Manhattan side of the river there is first an anchor span of 469½ feet, then a channel span of 1,182 feet, a span of 630 feet over the island, a second channel span of 984 feet, and finally an anchor span of 459 feet at Long Island City. The towers will rise a little over 300 feet above the water level. The total width of the bridge will be eighty-six feet, and the width between the centres of the trusses sixty feet. There will be an upper deck with two tracks for elevated trains and room for two more similar tracks when needed, and a lower deck carrying four trolley car tracks, a 36-foot central roadway and two 12-foot sidewalks. The structure is particularly noteworthy, apart from its magnitude, by the unique connections at the centres of the channel spans, where no suspended spans are used, and for the fact that 6,000 of the 45,000 tons of steel in the superstructure will be nickel steel.

The bridge was designed by Gustave Lindenthal while he was Commissioner of Bridges of New York. No work has yet been done on the superstructure, although the masonry has been partly erected.

The plans for the fourth bridge, of the suspension type, have not received, at the time of writing, the approval of all the authorities who must pass on them. This structure will be known as the Manhattan Bridge, and it will cross the river just north of the Brooklyn Bridge. It will have a main span of 1,470 feet and two land spans of 725 feet. The steel towers will rise about 320 feet above the water, and will carry four cables about 21 inches in diameter, hanging vertically and not cradled like those of the Brooklyn Bridge. There will be an upper deck with four tracks for elevated trains, and a lower deck 120 feet wide with four trolley car tracks, two 10½-foot sidewalks and a 35-foot roadway. The two stiffening trusses will be 24 feet deep and form a much less conspicuous feature of the bridge than those of the Williamsburg Bridge.

This bridge was designed by O. F. Nichols, chief engineer, and R. S. Buck, consulting engineer, of the Department of Bridges of the city.

FROM NEW YORK TO SCHENECTADY

After leaving the Grand Central Station on Forty-second Street, the train quickly plunges into a tunnel which continues for two miles. Half a mile beyond the upper entrance of the tunnel the train traverses a great steel viaduct with four tracks which extend from 109th Street to Mott Haven, the great four-track drawbridge over the Harlem, the heaviest in the world, being only a link in this viaduct. From this viaduct on the left-hand side an excellent view can be had of the upper end of Central Park, which is in the height of its beauty during the month of May. Conspicuous among the foliage stands the old Mount St. Vincent buildings, now one of the restaurants in the Park.

Immediately beyond the Mott Haven station the tracks of the New Haven and Harlem Railroads diverge to the right, and those of the New York Central follow the course of the Harlem River. After the train passes under McComb's Dam Bridge an excellent view may be had of the new speed-way on the other side of the river and of the Jumel mansion on the wooded heights just where the speed-way begins.

Seven miles out the train passes underneath High Bridge, which carries the old aqueduct from the mainland to Manhattan Island, and a few hundred feet further may be seen the terminus of the new aqueduct which here tunnels by a deep siphon 300 feet below the river bed.

The train now passes under the magnificent Washington Bridge, constructed between 1886 and 1890 at a cost of $2,700,000, with two central arches each of 510 feet span, and a roadway 150 feet above the level of the

Residence of Andrew Carnegie

river. The railway next makes a sharp curve at King's Bridge and soon reaches Spuyten Duyvil station on the Hudson River.

Hitherto the train has kept the left hand tracks on account of the Grand Central Station having been built when nearly all the railways in the country ran their trains on the left hand tracks. At Spuyten Duyvil the train now changes over to the right hand tracks, as is now usual in all the railways in the country.

After passing Spuyten Duyvil a fine view may be had

of the Palisades on the right bank of the Hudson. These are from 200 to 500 feet in height.

Fourteen miles from New York the new Mount St. Vincent Convent may be seen, among the buildings of which is Fonthill, formerly the home of Edwin Forest, the actor. Two miles farther Yonkers is passed, and between that city and Tarrytown glimpses may be caught of many splendid country houses, owned for the most part by rich business men of New York city.

The order of the stations after leaving Yonkers is Hastings, twenty miles from the Grand Central Station; Dobbs Ferry, twenty-one miles; Irvington, twenty-three miles, with "Sunnyside," Washington Irving's old residence covered with ivy; and Tarrytown, twenty-six miles. Between Irvington and Tarrytown lies Lyndehurst, formerly the Paulding Manor, and more recently the residence of the late Jay Gould.

Thirty miles out the train passes under Sing Sing prison by two short tunnels, the town of Ossining lying a mile farther up the river. Shortly afterwards the train passes over, by a new steel girder bridge, the Croton River, from which the supply of water for New York City is taken. The Hudson River is here at its widest, being four miles across, and this part of it is known as Haverstraw Bay.

Peekskill is passed at forty-two miles out, and the summer camp of the National Guard of the State of New York may be seen immediately after passing the town, while Dunderberg Mountain lies on the opposite side of the Hudson, rising to a height of 1,090 feet. Five miles farther the train penetrates Anthony's Nose, 1,230 feet high, by a tunnel seventy yards long, and three miles beyond Garrison's, the station opposite West Point, is passed, and Storm King, 1,530 feet in height, may be seen on the west side of the Hudson six miles above West Point.

At Fishkill Landing, fifty-nine miles from New York, the terminus of the New York and New England Railroad is passed, opposite Newburg and near the mouth of Mattewan Creek. Poughkeepsie, which is only a little more than half way between New York and Albany, is the next place of any importance. Beyond the town may be seen the fine new buildings of the State Insane Asylum. Vassar College, which is also situated here, is not visible from the train.

Between Poughkeepsie and Hudson lie many fine mansions, some expensively built in modern style and others which have been standing since colonial times though kept in excellent repair. Rhinecliff, near which the country residences of ex-Governor Morton and John Jacob Astor are situated, is passed at eighty-nine miles from New York, and on the opposite side of the river lie Rondout and Kingston.

Catskill Station, 111 miles out, has a ferry which runs across the river to the town of Catskill, which is the chief avenue of approach to the mountains of that name which are here prominent on the west side of the river. Hudson, 115 miles out, marks the head of ship navigation though steamboats go as high up as Troy.

East Albany is reached 142 miles from New York, where a good view of the State Capitol may be had from the left side of the train. Just before reaching Albany, the tracks of the New York Central cross the Hudson River. After leaving Albany the train ascends the steepest grade of the New York Central main line, after which it passes the large car shops at West Albany.

After passing the 158th mile stone, the first view of the Mohawk Valley is seen, the city of Schenectady lying in front and the works of the General Electric Company to the left.

NEW YORK EDISON COMPANY

New York Edison Company

THE New York Edison Company of to-day is the successor of the Edison Electric Illuminating Company of New York, the first corporation ever organized to do incandescent electric lighting on a permanent basis. The Illuminating Company was organized on December 17, 1880, as the New York licensee of the Edison Electric Light Company, in which was invested the ownership of Mr. Edison's patents relating to electric lighting. The Light Company was organized on October 16, 1878.

AREA COVERED AND SYSTEM EMPLOYED

The system for the supply of electric current covers practically the entire Island of Manhattan, containing 21.93 square miles and an estimated population of 1,900,000; and the Borough of The Bronx, having 40.65 square miles and a population of 625,000. The Boroughs of Brooklyn, Queens and Richmond, having 264.5 square miles and a population of 1,519,653, completing the City of New York, are not included in the system of The New York Edison Company.

The Edison three-wire system is employed, supplying low-tension, direct current, 120-240 volts, on Manhattan Island, and alternating current, three-phase, sixty-cycle, at 2,000

Isle of Safety at Fifth Avenue and 23d Street. The first in New York

volts and converted locally for distribution from a secondary network at 120-240 volts, is supplied in The Bronx. The Manhattan system is entirely under ground, the Bronx system, with slight exceptions, is overhead.

GENERATING AND SUB-STATIONS

The steam generating stations, which also distribute current locally are:

53-55-57 Duane street extending to Pearl street.

115-119 East 12th street.

45-47 West 26th street, extending to 27th street.

140th street and Rider avenue, Bronx.

The Waterside Station, occupying the entire block between First avenue and East River and 38th and 39th streets.

The sub-stations to which power is supplied exclusively by transmission from Waterside Station are:

11 Broadway (The Bowling Green).

39-43 Gold street.

200 Elm street.

96-98 Vandam street.

152 Clinton street.

32 Horatio street.

452 West 27th street.

117-119 West 39th street.

118-122 West 53d street.

123 East 83d street.

211 West 84th street.

128 East 121st street.

258 West 124th street.

Current is also supplied for conversion to the Bronx Station at 140th street and Rider avenue. Two new sub-stations are in the course of erection, one at 44-46 West 27th street, another at 167-169 West 107th street. With the exception of the Bowling Green plant, all of the stations supplying current to the Edison network, are the property of the New York Edison Company, and each, with rotaries and storage batteries, is fully equipped as a permanent centre of supply. The steam generating stations are also equipped in part with rotaries and storage batteries.

CAPACITY OF PLANTS

In aggregate ground area these properties, including the new Waterside, have 229,549 square feet, of which 39,500 square feet belong to the present Waterside Station. The steam generating stations, exclusive of the Waterside Station, aggregate 13,832 horsepower in boilers, 21,875 horse-power in engines, 14,600 kilowatts in generators, and 132,000 ampere hours, 135 volts, at a three hour discharge rate, in storage batteries. With the Waterside added, the rated boiler capacity is increased to 50,232 horse-power, the normal engine capacity to 82,375 horse-power, and that of the generators to 53,100 kilowatts. One of the two storage batteries in the Waterside Station, of 6,000 ampere hours, 135 volts at a 3 hour discharge rate, is devoted exclusively to the field excitation of the generators.

Standard post for Street Lighting developed by The New York Edison Company

The sub-stations, containing 7 motor generators—3 of 150 kilowatts and 4 of 500 kilowatts—and 57 rotary converters—2 of 400 kilowatts, 22 of 500 kilowatts and 33 of 1,000 kilowatts—have an aggregate capacity of 47,250 kilowatts. Four 2,000 kilowatt rotaries—the largest yet made—are in the course of installation. The 22 storage batteries in the substations have a total capacity of 44,000 amperes—132,000 ampere hours—at the three hour discharge rate, and it is estimated that at the end of the pres-

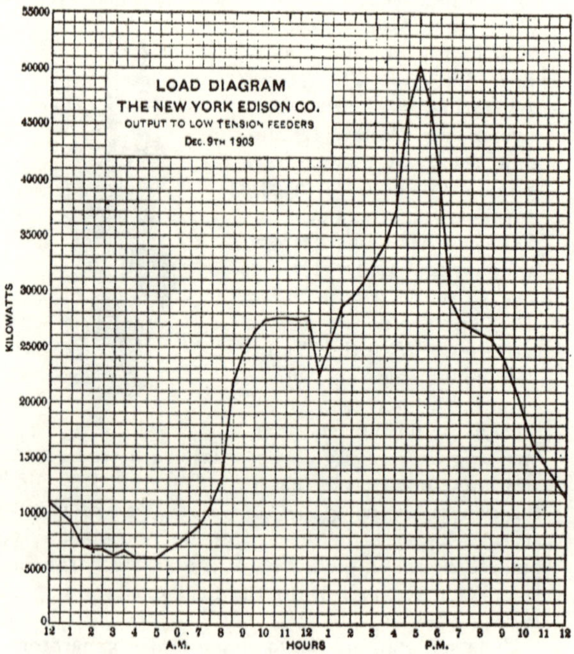
LOAD DIAGRAM
THE NEW YORK EDISON CO.
OUTPUT TO LOW TENSION FEEDERS
DEC. 9TH 1903
KILOWATTS
55000
50000
45000
40000
35000
30000
25000
20000
15000
10000
5000
0
12 1 2 3 4 5 6 7 8 9 10 11 12 1 2 3 4 5 6 7 8 9 10 11 12
A.M.
HOURS
P.M.

ent year, 1904, at the one hour rate, the aggregate battery capacity will equal twenty-five per cent. of the combined capacity of the various plants of the Company.

CURRENT GENERATED

For all the generating stations the maximum load on Manhattan Island during 1903 was 359,460 amperes, which at the average station pressure of 140 volts, equalled 50,600 kilowatts. Contrasted with this is the maximum load upon the Bronx station, which reached 1,540 kilowatts. It is expected that the Manhattan Island maximum of 1904 will approximate 425,000 amperes at 140 volts. The output of the generating stations on Manhattan Island during 1903 aggregated 131,000,000 kilowatt hours. The ratio of maximum load to connected installation was 36.6 per cent.; a percentage which has been decreasing from year to year with increased concentration in the generation of current and greater diversity in the purposes for which it has been employed. The average load on the Waterside station for the maximum day of 1903, December 24th, was 21,535 kilowatts, a load factor of 56 per cent. of the twenty-four hour capacity; the average for the month of December was 18,059 kilowatts, 47 per cent. of the capacity.

INSTALLATIONS SUPPLIED.

Current is supplied through more than 42,000 meters, to installations aggregating 1,507,342 incandescent lamps, 19,386 arc lamps, 85,072 horse-power in motors, and 2,000 kilowatts in customers' storage batteries, heating and cooking appliances and experimental apparatus. Of the arc lamps 3,126 are supplied for the municipality. The aggregate of these installations is the equivalent of 2,955,214 lamps of sixteen candles, each averaging 50 watts.

THE DISTRIBUTING SYSTEM

At the end of 1903 the underground system aggregated 334.67 miles of mains, 152.82 miles of low-tension feeders, and 107.44 miles of cable in the high-

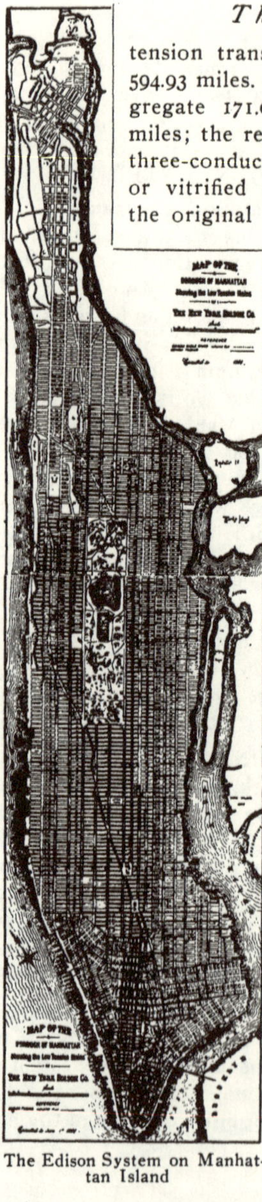

The Edison System on Manhattan Island

tension transmission system, a total of 594.93 miles. The Edison tube mains aggregate 171.06 miles; the feeders 65.52 miles; the remainder of the system is of three-conductor cable enclosed in iron or vitrified clay duct subways. When the original Pearl street station started there were 4½ miles of feeders and 10½ miles of mains—all two wire—in Edison tubing. The present tube system is maintained carefully, but otherwise in all new work cable is used exclusively, both for mains and feeders. Cable in iron ducts is also used for the service connections. The overhead system of the Bronx equals about 350 miles of overhead circuits.

All cable mains, single conductor, are stranded and each conductor has an area of 200,000 circular mils. In special instances, where the main acts as a tie between important points or where a large installation is to be served, larger sizes—350,000 circular mils and over — are used. Some single buildings have as many as four services, each having an area of 1,000,000 circular mils, or about one inch cross section. In such instances, several feeders will converge nearby on the local network, which will be tied together and strengthened in every possible way.

SERVICES

In all instances the services are brought into the building to be supplied at the expense of the Company. They terminate on or near the front wall in a switch, by which the entire supply may be controlled. Fusible safety devices are installed at this point to cut off the current instantaneously should trouble develop on wiring, fixtures or apparatus. In usual practice the meter will be placed near the service end, in a position insuring dryness and freedom from vibration, and otherwise favorable for the accurate measurement of current.

As the Edison tubes are supplied in lengths of about twenty feet—the width of a city lot—they are adapted to the convenient installation of an independent service connection for each building. The iron pipe and tile duct systems have handholes placed in the branch lines, accessible from the street by removing an iron cover, from which, placed short distances apart, building connections may be conveniently made.

NUMBER AND SIZES OF FEEDERS

The outside conductors—those of positive and negative polarity—of the Edison tubing reach a maximum size of 1,000,000 circular mils; the neutral conductor of these feeders has about one-third of this capacity. For other than tie purposes between stations, these feeders rarely exceed a length of one mile—under usual conditions about the greatest distance of economical supply at 240 volts pressure. The cable feeders are concentric, two-conductor, one for each polarity. The neutrals of these feeders are contained in independent cables of 2,000,000 circular mils, following the "tree" method, each being common to a number of feeders, thus providing at any

High tension gallery, 83rd Street Station

given point very much larger conducting capacity in the event of serious disturbance in the "balance" of the system. On the low-tension network there are

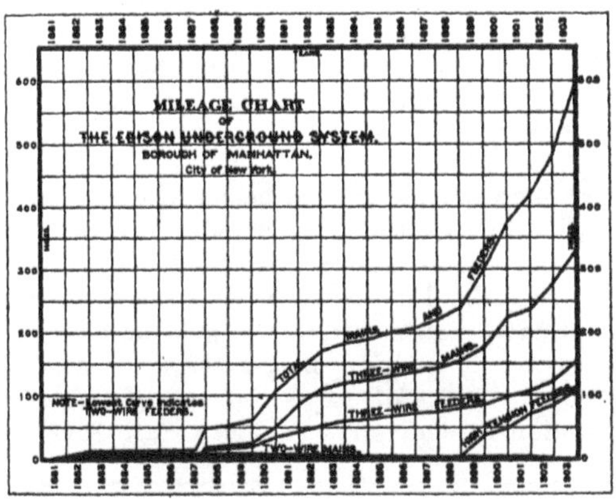

now 346 feeding points, and the number is rapidly increasing.

THE EDISON JUNCTION BOX

An important feature of the Edison underground system is the junction box. The box is of heavy cast iron, circular in form, with tube stubs at the bottom, the number and whether for mains or feeders being determined by the type of the box. It is enlosed with heavy iron covers, one tightly bolted down on a rubber gasket, thus making the interior water and air tight, the other lying loose on a suitable flange for the purpose of protecting the inner cover, maintaining the street level and supporting traffic. In the interior of the box are heavy copper rings one for each of the polarities, which connect by flexible cables with the conductors contained in the stubs—to which in turn mains and feeders radiating from the box are connected. All conductors make common connection with these rings, the positive and negative, through safety fuses, which, in the event of overload arising from accident or other cause, melt and sever the defective part from the general system. The neutral series of conductors is continuous without safety fuses, and is carefully "grounded" at each box. Thus the entire system of mains and feeders interlocks, and yet is fully protected at every point.

HIGH TENSION SYSTEM

The high tension system began in the latter part of 1896, when cables were installed for connecting a temporary substation, erected at the corner of 72nd street and Fifth avenue, with the high-tension generating station at 80th street and the East River. The second installation of this character was made in 1898, one series of cables connecting the generating station at 80th street and the East River with

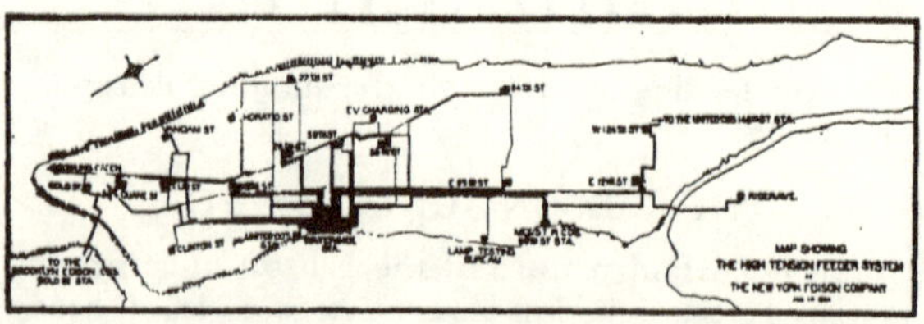

High tension feeder system in Manhattan Island

the substation on East 83rd street near Third avenue; another series was placed in the Broadway subways connecting the Duane-Pearl and the 39th street stations. This latter installation utilized, by inverting the rotaries, the switchboard current of the Duane street for assisting 39th street and 39th street current for assisting Duane street as each movement became necessary with the change in the loads upon the respective stations, there being some difference in the time between their maximum loads.

Two or more cables are now provided, by different subway routes through their entire length, between the Waterside station and each station and substation. In some instances the number of cables is greater, as for example, at Duane-Pearl street station there are 4 cables, and at 26th street there are 5, each connected with the Waterside by independent routes. There is also a general tie feeder which either loops or tees into all the stations and substations from Duane to 124th streets. This feeder may also be used to transmit high-tension current from one converting point to another, independently of the Waterside station. The high tension transmission system also extends into the generating station of the Metropolitan Street Railway Company at 96th street, and across the Brooklyn Bridge to the plants of the Edison Company of that Borough.

Each high-tension feeder begins practically at the automatic oil switches on the first floor of the operating gallery of the Waterside station; before this point is reached each pair of feeders is controlled by non-automatic oil selector switches, located on the second floor of the gallery. The selector switch enables the connection of the feeder on either of the two main busses of the station. The control of the oil switches of each feeder is concentrated on a feeder panel situated on the third floor of the operating gallery, upon which are also mounted the various indicating and recording instruments belonging to the feeder. Likewise for the terminals of the feeders at the converting stations there are feeder panels, upon which suitable switches have been placed for their control at that point.

INSULATION OF HIGH-TENSION FEEDERS

Rubber insulation was used for the first of the high-tension transmission cables, but in the more recent work paper has been used exclusively. The specifications for these cables were drawn to insure the best utilization of the subway ducts, and called for three conductors, each aggregating 250,000 circular mils and made up of thirty-seven strands of copper wire. The paper insulation is 5-32 of an inch around each conductor, and the outside insulating jacket is of the same thickness. The lead covering is 4-32 of an inch in thickness, and alloyed with from two to three per cent. of tin. The open spaces between the conductors are filled with dry jute saturated with an insulating compound to exclude air and moisture. It is required that after being laid in the subway, the insulation of the cable, including the joints, shall be 300 meghoms per mile at 60° Fahrenheit. In accordance with the rules of the Subway Company, each feeder is subjected to weekly tests. The capacity of each feeder is 250 amperes per phase at 6,600 volts.

THE SUBWAY SYSTEM

The subway system, which has been developed by separate corporations, extends through every important section of Manhattan Island. It consists of iron pipe or vitrified clay ducts, in groups of from two to thirty ducts, from 2.5 to 4 inches in diameter, buried in concrete. Manholes are provided at each street intersection, the distance apart being about 250 feet; on the trunk lines, passing through the long cross blocks, the intermediate distances may be a little in excess of 250 feet. Handholes serve the same purposes where the subway has more limited capacity. All cable joints are made in either manholes or handholes, where they may be easily cut for testing or repairs. The man holes are built of brick enclosed with double iron ventilated covers (the inner being locked), ventilated to prevent the accumulation of gas. Where passing through the manholes the cables are carefully stacked on iron racks fastened to the walls; those belonging to the high-tension system are covered with a wrapping of asbestos and galvanized steel tape, which affords protection from mechanical as well as electrical injury, the latter otherwise a possibility in the event of short circuit upon other cables.

Extending from the Waterside station there are

four independent routes of trunk subways, each containing from 20 to 30 ducts. Accident in one, however remote, by no chance can extend to the others. In addition to the Edison tube system, complete with junction boxes, there are 308.5 miles of subway aggregating 1,656.6 miles of single duct, 1,930 manholes and 15,715 handholes.

ROTARY STATIONS

If the five steam generating plants, with their rotary equipment, be included there are now 18 Edison rotary and storage battery substations in operation on Manhattan Island. Twelve of this number are entirely and permanently in the substation class, and receive all the power they convert from the high-tension feeders reaching out from the Waterside switchboard. The typical equipment of all is much the same, although their capacity varies with the needs of the neighborhood in which they are located. There is also some variety in the interior arrangement, but the practice more nearly standard, in that it more nearly meets the demands of economy, efficiency and convenience, is the location of cable vaults in the basement, of the rotaries and low-tension switchboard, including control of the end cell switches, on the first floor, of the statics, induction regulators, and high-tension switches on a mezzanine gallery, and the battery on a floor erected above. In some instances the batteries have been placed in the basement and the switchboards, high and low-tension, on a mezzanine floor extending across the front, and in others the rear, of the building. In the 26th Street Annex the rotaries occupy the basement, but here for other than an entrance or mezzanine floor, the room has been carried to the usual height of a second story.

The architectural treatment varies in detail, but is in harmony with the purpose of the structure. The space occupied varies from one to two city lots, from 20 to 40 feet in width, by the usual depth of

100 feet. An arrangement which combines a maximum of capacity in a minimum of space may not from an operating point of view be best, but in the development of the New York system it is necessary engineering. In the Annex of the 26th street station there are five rotaries, each of 1,000 kilowatts capacity, on a ground area of 16 by 100 feet, including passageways.

ROTARY STATION EQUIPMENT

In addition to the rotary converters and storage battery, the usual equipments of a substation includes:

Static transformers, air cooled.

Induction regulators for the rotaries.

A booster set for the batteries.

A direct current compensator for the three-wire system.

A switchboard in several panels, with devices for:

(a) High-tension feeder control.

(b) Low-tension feeder control.

(c) Rotary converter control.

(d) Battery control.

(e) Indicating, recording and synchronizing instruments.

(f) Busses, main and auxiliary.

COURSE OF CURRENT IN ROTARY STATION

Entering the station the three-phase, 6,600-volt, 25-cycle, alternating current is received upon high-tension feeder switches; from these connections are made, through the high-tension busses, with the selector oil switches, which permit any rotary to receive current from any feeder; from this point the current passes over duplex cables to the high-tension side of the static transformers. Here the pressure is reduced to 180 volts, alternating current. The secondary sides of the transformers are connected through the induction regulators with the alternating current collector rings of the rotaries. From the direct current side of the rotaries the path leads directly to the low-tension switchboard, where suitable switches provide that connection may be made with any one of three busses, each maintained at a different pressure, which supply the low-tension, direct current, feeders.

On the direct current side the normal pressure of the rotary converters is 270 volts, which may be raised or lowered 30 volts by the induction regulators. Any tendency toward unbalance on the three-wire system, which provides 120-240 volts at the services, is cared for by the battery and compensator.

STATIC TRANSFORMERS

All static transformers are air cooled and stand of sets of 3-200, 3-400 and 3-800 kilowatts respectively for the 500, 1,000 and 2,000 kilowatt converters. They are mounted on a platform containing the air duct for ventilation, in which, supplied by electric blowers from either end, or both ends, the air pressure is maintained at one-half or three-quarters of an ounce. The static transformers are wound for a ratio of transformation of 6,300 to 170 volts. The later types contain a thermometer, placed in the casing between the transformer coils, thus giving temperature indications.

Without undue heating the transformers will operate at 25 per cent. overload for three hours, or 50 per cent. overload for one hour, after a twenty-four hour run at full load. The efficiency of the 400 kilowatt type is 98 per cent., and the regulation is 1 per cent. The use of electrically operated alternating current switches on the transformer switchboard, controlled by small switches from the operating switchboard, results in shortening heavy cables and saving space on the operating board.

INDUCTION REGULATORS

Induction regulators permit a variation of from 150 to 190 volts in the pressure of the alternating current at the rotaries. Their secondary windings are conected between the secondary side of the transformers and the collector rings of the rotaries; their primary coils are wound on a rotor which by means of a small direct current or induction motor controlled from the operating switchboard can be turned through a given angle in either direction. These regulators have a capacity of 65 kilowatts for rotaries of 400 kilowatts and of 130 kilowatts for rotaries of 1,000 kilowatts. Standing upon the same platform they are cooled from the air ducts supplying the static transformers.

ROTARY CONVERTERS

The standard sizes of rotaries now used in the substations of the Company are 6-phase, 500, 1,000 and 2,000 kilowatts, respectively. They convert to

2000 Kw. Rotary

240-300 volts direct current; the speed is 500 kilowatts, 375 revolutions; 1,000 kilowatts, 187.5, and 2,000, 115 revolutions per minute. Some rotaries of 500 kilowatts, converting to 240-340 volts direct current, are used but mostly in the upper sections of the city, where a number of long feeders are still necessary. They are provided with induction regulators of 130 kilowatts capacity, which give them this unusual range of pressure.

The overload capacity of the rotaries is 25 per cent. for three hours, or 50 per cent. for one hour, after running continuously at their rated capacity. The efficiency of the 500 kilowatt type is 95.75 per cent.; of the 1,000 kilowatt type, 96.75 per cent., and of the 2,000 kilowatt type 97 per cent.

STORAGE BATTERIES

The recent standard storage battery equipment used by the Company consists of chloride accumulators furnished by the Electric Storage Battery Company of Philadelphia. Their type "H" cell is used, each cell containing 29 plates and being capable of discharging 500 amperes for 8 hours, 748 amperes for 5 hours, 1,120 amperes for 3 hours and 2,240 amperes for 1 hour. The plates are contained in wooden, lead-linked tanks, 48 inches high, 21 inches long and 34 inches wide. They contain 754 pounds of acid and, including the acid, the weight of each completed cell or tank is 2,492 pounds. There are 150 cells in each battery, 75 on the positive and a like number on the negative side; 20 cells on each side are connected to the end cell switches.

Battery room, Elm Street Station

A variety of floors have been used in the battery rooms. The latest installed have the regulation acid proofing of building paper and asphalt, followed by four or five inches of concrete, a thin layer of asphalt and building paper, and surfaced with vitrified tile three inches in thickness, in blocks, each block six inches square. The joints between the tile are very small and are filled with asphaltic pitch; the tile itself, which is made especially for this purpose, is almost absolutely non-absorbent.

It is now proposed to try experimentally the installation of a plain concrete floor treated with paraffine, applied under pressure at a temperature of 225 degrees Fahrenheit. It is claimed that by this process the pores of the concrete can be completely filled to a depth of at least three inches. This method of construction, though as yet untried, promises to be more satisfactory than any of the others previously followed.

STATIONS

In the construction of any building on Manhattan Island, the most economical utilization of every square inch of available space is the pressing problem upon architect or engineer. Being entirely surrounded by water, the island offers opportunity for extension in only one direction. Preceding the advent of high-tension transmission the advantages derived from placing the generating point in the heart of the district supplied more than offset the advantages incidental to a position on the water front; hence the vertical station, with engines and generators at the foundations, and boilers and coal bunkers high above the level of the street.

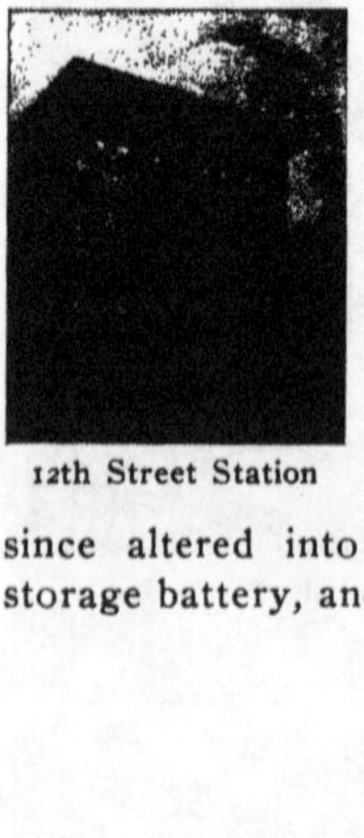

12th Street Station

Three plants of this vertical type have been created—Thirty-ninth street, placed in service Thanksgiving Day, 1888, and since altered into a substation with rotaries and storage battery, and facilities for the arc lamp, meter and other departments; the Twenty-sixth street station, placed in commission on Christmas Day of the same year, 1888, still in operation, greatly enlarged, and third in size of the Company's plants; and the Duane-Pearl street plant, exceeded in size only by the Waterside, and containing the general, executive and administrative offices of the Company. The Twelfth Street station, which followed, was erected on horizontal lines, the engine and boiler rooms occupying parallel lots of ground, but it was not intended that these works, somewhat to one side of the electrical centre, and in a territory promising only limited growth for a considerable period, should be more than an annex to the larger plants.

Duane-Pearl Street Station

DUANE-PEARL STREET

The Duane-Pearl street property, 100 feet in width by 200 feet in depth, facing through its entire length on Elm street, was purchased late in July, 1890. Very little time was lost, for on the first day of August began the removal of the old buildings and the excavations for the new station.

Current was first supplied from this site on November 1, 1890, from a plant temporarily placed in the basement of a small brick building still standing on the Company's property at the corner of Duane and Elm streets. On May 1, 1891, coincident with the abandonment of the small annex station in the cellar of a real estate office on Liberty street, previously assisting the original Pearl street station and the first of its kind in existence, a temporary installation, erected on the concrete floor of the permanent construction, was placed in commission, and continued in operation until it was superseded by the permanent equipment.

View of operating room, Elm Street, looking north

The width of the building is 74 feet, with a depth of 200 feet, extending from street to street. Both façades, designed to indicate the purpose of the plant, and even the part of the work to which the individual floors are devoted, are substantially alike. The first three floors are of dressed granite, extending through the thickness of the walls; above a buff colored pressed brick is used, with the slight attempt at ornamentation of placing at various points multipolar generators and incandescent lamps, in terra cotta. Above each entrance are four decorated panels upon which have been placed the profiles of Volta, Ohm, Ampere and Watt, on the Pearl street façade, and Edison, Franklin, Henry and Morse, on Duane street. The side walls of the building, having a thickness at the base of nearly five feet, are almost without ornamentation; the iron work was the heaviest and most substantial that to that time had been placed in any building in New York. The characteristic stacks, 150 feet high above the grate

Operating room, Elm Street, showing high tension gallery

bars and extending 75 feet above the roof, at the centre of the building, with a spiral iron stairway between leading to an ornamental gallery around the top of each, have long been known as the "Heavenly Twins." Their diameter is 14 feet, or 12½ feet clear of the inside brick lining.

Illustrated by a cross section view the interior arrangements may be readily understood. The basement, between the engine foundations, is occupied largely by the cable runs with feeder vaults at each end; on the first floor are the engines, generators and the operating switchboard; on the second floor, extending from the first through to the boiler room, is a space devoted to the steam piping, steam separators and water heaters; the shops and storeroom were placed originally on this floor, but they have been recently superseded by two storage batteries of large size, with booster and other auxiliary apparatus; the available space on the third floor is occupied by the ash runways and the boiler feed pumps, and on the floor above are the boilers; the sixth is a mezzanine floor, which, with the fifth floor, is occupied by the coal bunkers of 2,000 tons capacity, the main smoke flues, water storage tanks and coal conveyor. The seventh and eighth floors are occupied by the General Offices, which, because of overcrowding, also extend to the top floor of the large building on the opposite side of Duane street. In addition to the storage battery the present generating equipment consists of 9,133 rated horse-power in boilers, 11,800 horse-power in

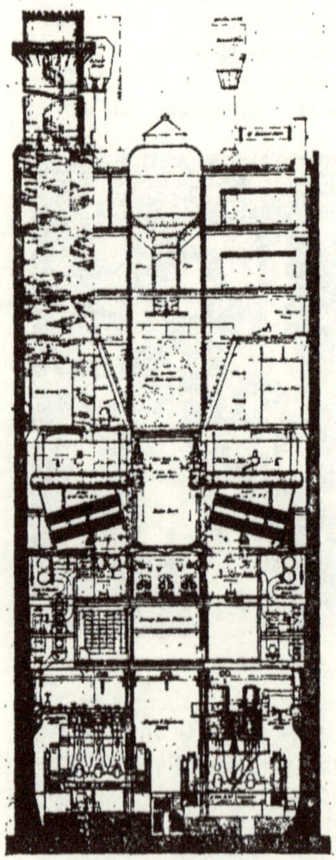

Sectional view,
Duane-Pearl Street Station

steam engines—three of 2,500, two of 1,250, and three of 600 horse-power—and 7,600 kilowatts in generators, which, two to each engine, correspond with the engine sizes. Each of the two storage batteries has a capacity of 6,000 ampere hours, 135 volts, at the three hour discharge rate, and the rotary converters have 4,000 kilowatts of capacity. Three 2,000 kilowatt rotaries are now in course of installation. The capacity of this plant at the end of 1904 will closely approximate 17,000 kilowatts.

At the time the engines of 2,500 horse-power, developed on the specifications of the Company's engineers, were the largest that had been constructed for work of this nature. This size was reached in three steps; the original plans contemplated an engine of 600 horse-power, and three of that size were installed; before the last one was ordered, however, it became evident that with some rearrangement of the parts, the floor space allotted was sufficient for an engine of 1,250 horse-power; and with the completion of this size it was found that by still further concentration, particularly in the rearrangement of the cylinders and the use of a two crank, four cylinder, quadruple expansion engine, the allotment of ground space would be sufficient for the larger size of 2,500 horse-power. It was in connection with these engines that the Van Vleck disconnective feature was designed, permitting the use of the unit at proportionally reduced capacity in the event of injury to any one of the several cylinders.

2500 Horsepower—Duane-Pearl Street Station

The multipolar generators of this station were also built upon the initiative taken by this Company, and were the first generators of this class to be manufactured or installed in connection with American central stations.

The switchboard, like the remainder of the plant, is a marvel in the economical use of space. At the outer edge are the main switches, field regulators and indicating instruments for the generators; behind on the face and rear of the board are the main busses and various connections, indicating instruments, and switches for the feeders radiating from the station. It was while working upon the problem of this board, searching for means of condensing and economizing in space, that Mr. Van Vleck designed the edgewise system, by which the movement of the controlling apparatus was changed from horizontal to vertical, resulting in what is now known as the Van Vleck Edgewise System. The Donshea attachment for switch interlocking, thus insuring absolute continuity in the field circuit while the generators are in operation, was also designed and first utilized in connection with this board. A feature in the operation of the switchboard is that the movement of any of the apparatus is in line with the result that will follow in the generation of current. Thus, in raising the main switch, the generator is connected, in lowering it, it is disconnected; in raising the handle of the rheostat, the pressure is raised correspondingly, in lowering the handle, the pressure falls; the upward swing of the volt or ampere meter indicates rising pressure or increasing current, the downward swing the reverse. Another feature is found in the indicating instruments which hang side by side; the scales and indicators are in alignment so that any variation from the standard pressure, or any change in the output of the generators or upon the feeders, is visible by glancing along the faces of the instruments. Any deviation from the common line indicates the extent to which the voltage or output varies from one point to another.

Some idea of the economy of the space that has been accomplished in this board may be obtained in the fact that providing for the present and ultimate capacity of the Duane-Pearl street works, it is only 40 feet in length.

THE WATERSIDE STATION

Originally the Edison plans for the supply of New York city contemplated as many as thirty-six independent districts and steam generating stations, south of Fifty-ninth street.

The limited range of low-tension distribution has been recognized as one of the disadvantages of the system, but the advantages, commercial and technical, were found sufficiently offsetting to justify the development of the distributing system of the New York Company on low-tension and direct current lines. The objections to steam generating stations in so many small districts, in both investment and operation, are too apparent to require explanation, yet in the absence of polyphase transmission they must have been brought into existence as, with the growth of the service, stations became cheaper than transportation —if the term may be used.

It is fortunate, therefore, that a system was in the meantime developed which, permitting the continuance of small centres of direct current distribution, has, by the simple provision of transforming apparatus, made possible the concentration of the great Waterside Station, economically located miles away from the point of use.

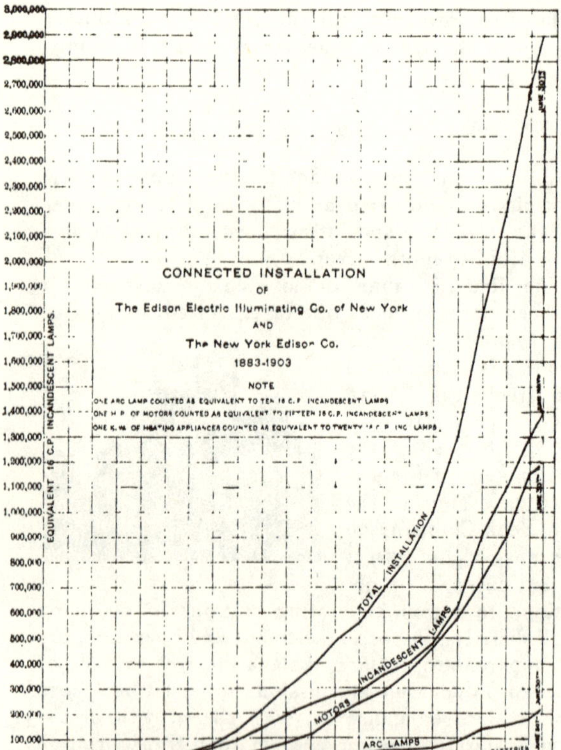
CONNECTED INSTALLATION
OF
The Edison Electric Illuminating Co. of New York
AND
The New York Edison Co.
1883-1903
NOTE
ONE ARC LAMP COUNTED AS EQUIVALENT TO TEN 16 C.P. INCANDESCENT LAMPS
ONE H. P. OF MOTORS COUNTED AS EQUIVALENT TO FIFTEEN 16 C.P. INCANDESCENT LAMPS
ONE K. W. OF HEATING APPLIANCES COUNTED AS EQUIVALENT TO TWENTY 16 C. P. INC. LAMPS.
EQUIVALENT 16 C.P. INCANDESCENT LAMPS.
TOTAL INSTALLATION
INCANDESCENT LAMPS
MOTORS
ARC LAMPS
HEATING APPLIANCES AND STORAGE BATTERIES

High-tension polyphase transmission, in combination with rotaries or motor generators, made the Waterside Station technically possible; the addition of storage battery auxiliaries, distributed over many points of supply, emphasized its commercial possibilities. It is not too much to say that in a lighting and power service of this nature, where reliability and continuity of supply are of paramount importance, the storage battery is one of the essential features of the scheme as a whole.

This system being available, the rapid growth of the Company permitted the Waterside Station quickly to take final shape. The new station as constructed occupies the entire block situated between 38th and 39th streets, First avenue and the East River, and when fully equipped as planned will have a capacity in excess of 100,000 horse-power. The property is 272½ by 197½ feet, and is located 160 feet from the edge of the water, of which 125 feet are occupied by a new marginal street. Tunnels for conveying condensing water from and to the river pass under this street. At the river edge a massive granite bulkhead has been constructed, upon which are mounted the coal receiving towers and ash pocket for the station.

THE BUILDING

Romanesque in design, the building is Renaissance in treatment. The general effect of massiveness is due to simplicity in lines and boldness in detail. The design consists of a base, shaft and capital, the base being formed by a massive rock-faced granite water table containing small openings of deep reveal. The effect of the shaft is secured by continuous vertical arched openings with a light filling of iron windows and spandrels, the smaller rectangular openings and the cornices over them forming a crown to the general design. The walls above the granite water table are of two shades of buff brick with belt and sill courses of red sandstone, and a cornice of terra cotta.

Here it is intended that the bulk of the current required for the New York system shall be generated. The steam stations still in existence, also equipped with rotaries and storage batteries, will operate during the "peak" hours of the year, and thus the Waterside Station will be relieved of that portion of the load. Storage batteries will also assist in leveling

The Waterside Station, from First Avenue

the peak on the Waterside plant. It is not suggested that the load of this station is or can be represented by a straight line, but rather that the advantages of this combination, together with the greater concentration in generation, the wider field of supply and the unprecedented variety of service, result in much closer relation than is usual between average and maximum demands.

BOILER SECTION

That portion assigned to the boiler plant, 76 feet in width, is plainly indicated in the architectural treatment of the façade. The basement of this section contains the flues, feed pumps, storage tanks and other auxiliary apparatus. The first and second floors are occupied by the boilers, above which are the coal bunkers, 275 feet in length and capable of holding 10,000 tons. Four stacks of steel construction, 17 feet in internal diameter, and 196 feet in height above the lower grates, surmount the entire section. They are lined with fire brick, eight inches in thickness, one-third, and with red brick the remainder of their height. The brick lining is in twenty-four vertical sections, each resting upon angle irons riveted to the shell of the stacks, between which and the brick there is an air gap of four inches.

BOILERS

There are 56 boilers of the water tube type, each 18 feet long with 6,500 square feet of heating surface, and rated at 650 horse-power. As the engines require only 12.5 pounds of steam per indicated horse-power, the boilers easily develop enough steam to meet all their requirements. As divided there are 28 boilers on each floor, standing in two rows, in batteries of two each, facing an aisle 30 feet wide. They are designed for 225 pounds, and now operate at 200 pounds pressure per square inch. There is a complete installation of blowers of large capacity for forced draft, and 20 of the boilers are provided with automatic stokers with the necessary auxiliary apparatus for their operation. The remainder, in which anthracite coal is used exclusively, are fired by hand.

COAL AND ASHES

Coal is received and ashes are discharged at the river front. For the coal there are two steel towers on the bulkhead, both equipped with clam-shell buckets, each of 1½ tons capacity. By means of these buckets the coal is taken from the boats, moored alongside, lifted to and deposited in large receiving hoppers, one on each tower. From the receiving hoppers, by means of a belt conveyer, the coal passes to weighing hoppers at the base of the inclined bridge, upon which there is a bucket conveyer for lifting the coal to and depositing it in the bunkers at the top of the boiler section. Beneath the receiving hopper is a crusher for reducing the soft coal, or in cold weather frozen lumps of hard coal to a uniform size. The crusher is provided with

Sectional view of the Waterside Station

a by-pass reached through a screen so that coal of less than a pre-determined size does not go through it. The conveyer over the bunkers is provided with tripping devices, causing the buckets automatically to turn and deposit their load at any desired point. Electric power operates the crushers as well as the conveyers. Either hard or soft coal may be received, and to safeguard against interruption in the supply, the entire plant is in duplicate. From barge to bunker the hourly capacity is 150 tons.

As has been stated, ashes are discharged at the river front. The ashes and the deposits from the combustion chambers fall through chutes from the floors above to cars in the basement of the boiler section, by which they are transported to a receiving hopper. From this point they are carried by a bucket conveyor to a receiving pocket on the bulkhead, constructed entirely of steel and concrete, and providing for nearly a week's accumulation of ashes. The scows, which carry from two to three days' ac-

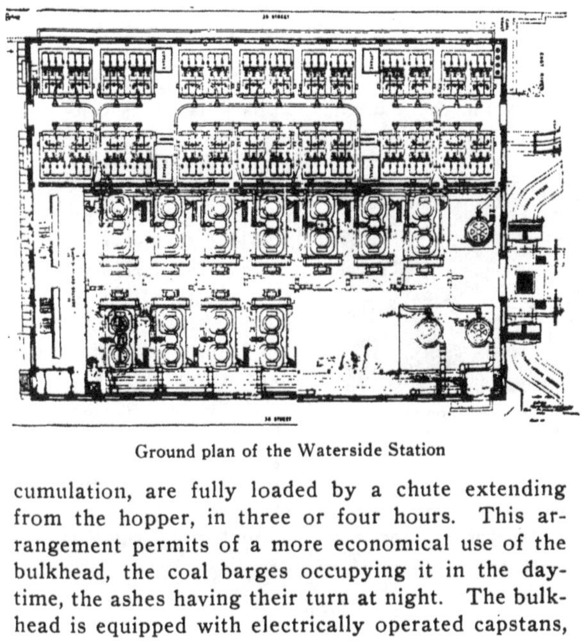

Ground plan of the Waterside Station

cumulation, are fully loaded by a chute extending from the hopper, in three or four hours. This arrangement permits of a more economical use of the bulkhead, the coal barges occupying it in the daytime, the ashes having their turn at night. The bulkhead is equipped with electrically operated capstans, and other conveniences for handling the boats.

THE MAIN OPERATING ROOM

The engine or main operating room is 115 feet wide, 267 feet 10 inches long and to the monitor at the centre of the roof nearly 125 feet in height. Running the entire length of the south side are five galleries devoted to the offices, shops and storerooms of the plant; on the westerly side, four stories high, enclosed in glass, are the electrical operating galleries.

Sixty-eight feet above the floor are the tracks supporting two travelling cranes, one having a lifting capacity of 50, the other of 25 tons, and both a span of 98 feet. In addition to their main hoists, each crane is equipped with a whip hoist of five tons. These cranes travel, fully loaded, at a speed of 150 feet, and lightly loaded 200 feet a minute. The hoist speed, fully loaded, is 25 and lightly loaded 40 feet a minute.

In the basement under the operating room are the condensers and their auxiliary apparatus, ducts containing the cables from the generators to the operating galleries, the feeder runs to the cable vaults under the sidewalk on the First avenue side, and two standard batteries, one for the local district service, the other insuring constant potential on the field excitation bus, occupy a section extending the length of the building on the 38th street side. The oil filters and pumps are also located in the basement. Below the level of the basement floor are the condensing tunnels leading to and from the East River.

A section of the condensing tunnel construction

ENGINES

It was intended originally to install 16 engines of the vertical marine type, three crank, with one high and two low pressure cylinders. The development of the steam turbine in the larger units has made it seem desirable to modify these plans somewhat, and at least a portion of the equipment will consist of turbines and their auxiliary apparatus.

In diameter the high pressure cylinder of each

engine is 43½ inches, the low pressure 78½ inches. The stroke is 60 inches. With 175 pounds of steam pressure at the throttle, 27 inches of vacuum and 75 revolutions per minute, at the most economical load the engines indicate between 5,200 and 5,500 horse-power. They are capable, however, of a sustained load of 8,000 horse-power, and of an ultimate capacity, at the maximum cut-off of five-eighths of the stroke, of 10,000 horse-power. The engine shaft, which is hollow, is nearly 30 inches in diameter; to give a uniform turning motion, the three cranks are set at angles of 101, 126 and 122 degrees; the flywheel weighs 90,000 pounds, and is twenty-three feet in diameter. The load or speed of any engine when running in multiple can be varied by the adjustment of the governor mechanism by an electric motor controlled from the operating gallery. In the event of accident there are emergency valves which permit the engine to be shut down immediately; should the speed become excessive, there is a centrifugal device in the crank which, automatically tripping a pilot valve, closes the large valve in the main steam line.

8000 Horse-power at the Waterside Station

The main steam pipes to the engines are 16 inches and the exhaust outlets between each low pressure cylinder and the condensers are 26 inches in diameter; each condenser contains 3,700 tubes three-quarters of an inch in diameter, and has an aggregate cooling surface of 9,200 square feet. Oil is supplied from a reservoir which, through a system of piping, connects with each oil cup and bearing. Between 15,000 and 20,000 gallons of oil are in constant circulation through this system. The oil runs from the collecting pans of the engines to oil filters in the basement, from which it is pumped to the reservoir at the top of the building.

GENERATORS

Directly connected to the engines the generators are of the revolving field type. They are rated for continuous operation at 3,500 kilowatts, 307 amperes per phase at 6,600 volts; they possess an overload rating for three hours of 400 amperes, and for short

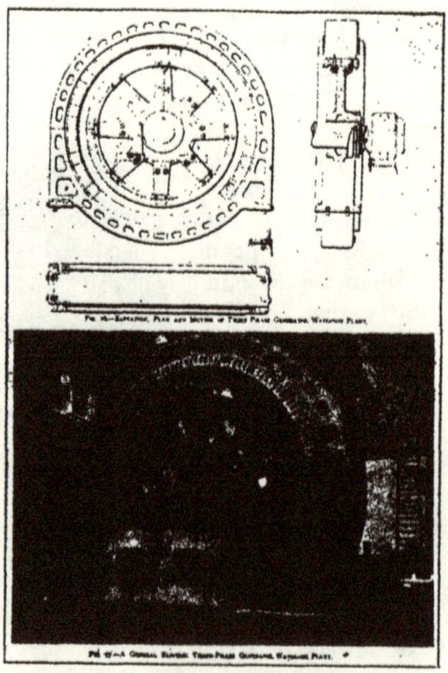

One of the Waterside Generators

periods are capable of sustaining an overload which is limited only by the capacity of the engine. Three phase current at 6,600 volts is generated; each unit has forty poles, and running at seventy-five revolutions per minute gives a frequency of twenty-five cycles per second.

The spider of the revolving field is keyed to the engine shaft and bolted to the hub of the fly-wheel. Thus upon occasion of excessive strain on the generators, the flywheel energy is transmitted directly to the spider, lessening the shearing effect upon the keys. The field windings are of copper ribbon set on edge, and give full excitation at 220 volts, making it possible to take the exciting current from the local low-tension switchboard, as well as from the storage battery and motor driven exciter set. The field current passes through two iron collector rings—iron because of its superior wearing qualities—for which carbon brushes give ample carrying capacity. The armature is built of laminated steel and the armature coils are form wound, placed in slots and held there by wooden wedges dove-tailed into the outer edges. The revolving field weighs 130,000 pounds, and the armature 125,000 pounds, a total weight for each generator of 255,000 pounds.

By test the actual efficiency for one-quarter overload is 97.2 per cent.; full load 97 per cent.; three-quarter 96½ per cent., and half load 95½ per cent. The regulation is such that if the full load be suddenly thrown off, the rise in the voltage of the generator will not exceed 8 per cent.

FIELD EXCITATION

Each exciter set, of which there are three in the station and a fourth in course of installation, consists of a 225 horse-power, 6,600-volt, 3-phase, induction motor, directly connected to a 150-kilowatt, direct current, four-pole generator giving a potential of 200-280 volts. A storage battery, occupying the vault on the 38th street side of the building, insures absolute continuity in the exciting current. The battery alone is capable of exciting the fields of sixteen generators for one hour.

THE SWITCHBOARD

The controlling, indicating and recording features of the electrical equipment of the station are confined to a series of galleries occupying the entire westerly end of the structure. On the main floor are the motor driven exciters, their switchboards and the controlling and indicating devices for the supply, control and record of the low-tension direct current distributed locally from this station. The rheostats of the exciter sets are located on a mezzanine gallery directly beneath.

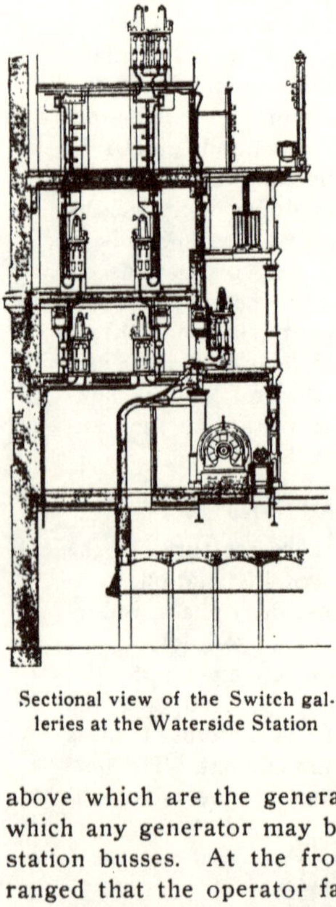

Sectional view of the Switch galleries at the Waterside Station

On the first gallery are the automatic oil switches controlling the feeders, and the transformers for the operation of their indicating and recording instruments. The main oil switches controlling the generators are also on this gallery. On the gallery above, the second, are the group selector switches controlling groups of two feeders, by which any group may be placed on either of the busses of the station, and the field rheostats of the generators.

The gallery above, the third, is the main operating gallery of the station. At the rear is the bus house, above which are the generator selector switches, by which any generator may be placed on either of the station busses. At the front of the gallery, so arranged that the operator faces and has in full view the machinery of the station, are the various switches and indicating and recording instruments incidental to and essential for the operation of the generators. Each generator is controlled from a pedestal upon which are mounted the controlling switches and apparatus, directly above which is a vertical panel containing all the instruments relating to that generator. The instruments on each generator panel consist of a recording wattmeter, giving a summation of the output of the generator; a voltmeter, two ampere meters, an indicating wattmeter, a field ammeter, a power factor indicator, a synchronizing lamp, and the signal lamp connected to the overload relay. There are also illuminated signals by which orders are transmitted from the operator on the gallery to the engineer in charge, or *vice versa.*

Above the operator's desk, in the centre of the gallery, are located the frequency indicators, station voltmeters and synchronizers. At this point are also located the ground detectors, which by means of relay circuits cause a bell to ring should a ground develop at any point on the system. Behind the operator, at the rear of this gallery, each feeder is provided with a vertical panel, upon which are mounted the switches controlling the motors operating the feeder switches, a recording wattmeter, three ampere meters, a power factor indicator and the time limit overload relays.

At one side of this gallery is stationed the system operator, upon whom depends the distribution of current between all the stations and substations of the Company in conjunction with the Waterside plant. Upon him rests also the responsibility for having adequate machinery in operation for any emergency, not only at the Waterside, but in all the other stations. The position he holds has been likened to that of a train despatcher on a great railroad system.

STEAM TURBINES

Three turbines of the Curtis type, manufactured by the General Electric Company, each having a capacity of 5,000 kilowatts, are now in course of installation, and will be in service in time to assist in the

Fall demands of this year. One of these turbines is now in position.

The first turbine is of the two stage type, with eight rows of revolving blades and six rows of guide blades, divided equally between the stages. The second and third turbines will be of the four stage variety, having four sets of expanding nozzles and

Curtis Steam Turbine—5000 Kw.

four revolving units, each unit containing two lines of blades, with one line of guide blades between them.

In the first type the steam is partially expanded at the first set of nozzles, flows through the four revolving blades and three guide blades of the first wheel, and expansion is then finished in the nozzles between the stages. The steam then passes through the four revolving blades and three guide blades of the second stage into the exhaust chamber in the base of the turbine, and there passes into the condenser, standing beside the turbine and of about the same height.

Communication is established between the turbine and the condenser by a rectangular pipe, having an area of about fifty-two square feet. The condenser is of the Wainwright, even flow, type, and is capable, while condensing 120,000 pounds of steam per hour, of maintaining a vacuum of one pound absolute, with the condensing water at 70 degrees Fahrenheit. Instead of the customary glands and packing, the tubes of the condenser are expanded into the tube plates—a marked difference from the usual type.

Suitable dry and wet air pumps and a centrifugal circulating pump have been provided for this portion of the station equipment. The oiling system for the turbines is provided with three pumps—two steam and one electric—either one of which is large enough for all the turbines. The required degree of oil pressure will be maintained by means of an accumulator.

THE NEW WATERSIDE STATION

So rapid has been the growth of the Company that it has become necessary to provide a new Waterside Station, and plans have been completed to occupy the block to the north of the present plant, bounded by 39th and 40th streets, First avenue and the East River. It is expected that a part of the new station will be in operation during the latter part of 1905.

As planned, the building will be trapezoidal in shape, having a frontage on First avenue of 197 feet 6 inches, on 39th street of 336 feet and 347 feet 2 inches on 40th street. The river end will be of the same length as the First avenue façade.

The building will be divided into a boiler house 123 feet 8 inches wide and a main operating or generating room 73 feet 10 inches wide. The boiler house will stand on the southerly half of the block, adjoining the corresponding section of the present station, the operating room occupying the northern portion of the property. The north side of the operating room will be occupied by the switch galleries, each about 15 feet wide, one above the other; the operating gallery will be at the west side of this room.

The boiler plant will consist of ninety-six 650 horsepower water tube boilers divided between two floors, 48 boilers to each, arranged on what may be termed the cross fire-room principle, in rows of three batteries, each battery being composed of two boilers. Two of these rows will be placed back to back, the flues, uptakes and one stack being common to each group of 24 boilers thus formed. In the completed plant there will be four such units and four stacks, the stacks having an internal diameter of 21 feet 6 inches, and rising 300 feet above the level of the lower grates—thus exceeding considerably the height of the present stacks.

Suitable forced draft apparatus will be provided. The piping from each boiler group will be arranged on the unit system, with cross connections available in case of accident. It is intended that the boilers shall be operated at 200 pounds pressure, and provided with superheaters capable of adding 150 degrees Fahrenheit to the temperature of the saturated steam.

Either vertical or horizontal turbines or reciprocating engines may be used in the operating room. The machinery ultimately installed at its normal rating will have a capacity of not less than 100,000 kilowatts. The switchboard arrangements of the new station will follow generally those of the present Waterside plant.

COAL STORAGE

To avoid the possibility of interruption in the coal supply, very large storage capacity has been provided. Labor questions at the mines or on the railroads, or storm blockades in transportation over the railroads, the water, or the city's streets, are possible causes of interruption. To avoid these at each step in its growth the Company has made what seemed to be adequate provision for coal storage. The original Pearl street station had capacity for

Coal Storage at Shady Side, showing the loading sweep

about fifty tons in sidewalk vaults, equalling about a week's supply. The stations next constructed, 26th and 39th streets, had bunkers capable of holding about one thousand tons. At Duane-Pearl streets the bunkers hold two thousand tons and at the Waterside ten thousand tons—about twelve days' supply.

Before the recent coal strike, during which large cargoes were imported from England—one alone containing 8,525 tons, the largest to that time ever crossing the Atlantic—it became evident that this provision was inadequate to give the desired degree of insurance to the New York service. Consequently in 1903 a large tract of land known as Shadyside, on the New Jersey shore of the Hudson River opposite the Grant Monument on Riverside Drive, Manhattan, was purchased and thoroughly equipped with coal handling machinery.

Coal may be delivered on the property by either

boat or rail and automatically placed in the storage piles. In the stations buckets on endless chains lift the coal to the bunkers on the upper floors and it is delivered at the doors of the boilers by gravity; in this storage yard it is lifted on enormous trusses and by gravity takes place on the pile, where it remains until needed for use.

The arc over each of the hard coal piles consists of two bow string trusses having a span of 340 feet and a height of 97 feet to the hinge. Coal is carried from the receiving hopper up on the side of the truss by a flight conveyer, and by adjustable openings in the track falls on the pile at any point desired between base and apex. Soft coal is stored by means of a bridge conveyer having a span of 220 feet and a run of 340 feet; it is taken from the boats by clam-shell buckets suspended and operated from a receiving tower at the bulkhead end of the bridge. The receiving hopper for hard coal is filled directly from cars standing on the tracks over it, or if delivered by boat, by a rubber belt conveyer, upon which crane buckets discharge their contents. As all deliveries from this yard must be made by water, this belt also serves to reload the hard coal from the receiving hopper at the piles to the point at which it is discharged into the vessels at the bulkhead.

As in storing, so in removing, the coal is handled entirely by mechanical means. The soft coal is picked up with clam-shell buckets, each holding about one and one-half tons, and deposited in the conveying barge anchored at the bulkhead. The hard coal is removed from the piles by a sweep supported on circular tracks passing under each pile. The sweep is moved to the inside base of the pile and, held there, the coal falls upon a scrape conveyer which carries it to a receiving hopper located at the hinge of the sweep, from which, by means of a screw conveyer standing at an angle, it is deposited upon the belt conveyer which, as stated above, runs to the bulkhead, several hundred feet away.

The present provision is for two piles of hard and one of soft coal. Each hard coal pile contains 60,000 tons without trimming; with trimming, the joint capacity of both piles may be increased to 160,000 tons. The soft coal pile stores between 30,000 and 50,000 tons.

Soft coal, it is expected, will remain on storage only for short periods, but the hard coal may remain indefinitely, its use being dependent upon the conditions of the market. At some seasons it may seem desirable to purchase and store in quantities, at others to cease storage and even to draw upon the reserves, though there be at the time no interruption in the market supply.

ELECTRIC AUTOMOBILING AND TRUCKING

A very interesting feature of recent and considerable growth is the Company's electric automobile service now including about twenty-five vehicles for Superintendents, delivery and trucking. The service is much faster than horses, perhaps almost twice as fast, and the cost is no greater, if as great. With the Superintendents' conveyances a higher degree of oversight is given important work than would be otherwise possible; the delivery wagons enable the distribution of supplies much more satisfactorily, and in a manner which presents a more pleasing appearance to the public; the electric trucks are capable of rapidly transporting loads as heavy as five tons. In addition to these advantages there is no doubt but that these vehicles possess considerable advertising value, and that they have materially aided in increasing the use of business automobiles propelled electrically on the streets of New York city.

One of the most interesting and useful applications of electric power, in connection with the transportation of heavy loads, is the design of a five ton truck by the New York Edison Company, upon which is mounted an electric windlass for drawing cables into the ducts of the subway, an electric pump for removing water from the manholes, and a switchboard which enables the battery to be charged while the truck is standing on the street. In addition to providing means for quickly drawing cables into the subways, this truck transports them on the reels from the storeyard to the point of use. The Company maintains a thoroughly equipped garage for this service, including storage room and repair shops.

THE METER

All of the direct current now supplied by the Company is registered on Thomson recording watt meters. More than 42,000 of these meters are in constant use. The alternating current in the outlying districts is supplied through mechanical watt hour meters of the induction type, of which there are about 2,000 in use.

All meters are tested before leaving the meter room and again within a month after their installation. After this each meter is tested annually, though to insure a high state of accuracy meters of the larger sizes are tested every two or three months. Tests are also made, at the expense of the Company, upon the complaint of a customer that his bills appear excessive. The number of meter tests made in 1903 was 46,924, somewhat in excess of the total number of meters installed.

The Meter Department is administered from the upper portion of the former 39th street station, where elaborate facilities have been provided both for direct and alternating current work. The alternating current is supplied at 60 and 25 cycles, this variety being sufficient for the purposes of the Company, and the direct current connections are capable of supplying as many as 1,000 amperes at the test board at either 120 or 240 volts. A small storage battery of 140 cells, 70 to a side, is also connected with the testing board; end cells, with suitable connections, give any required voltage. The equipment of the Department also includes a bank of 1,100 lamps giving the required current flow for meters of the various sizes.

The Department's equipment also includes apparatus for testing the various parts of meters, including the armatures, magnets and jewels. A complete repair shop is maintained and spare parts for each type of meter are kept in stock for immediate use, thus it is possible to make hurried repairs, without returning the meters to the factory.

RATING OF METERS

Owing to the fact that these meters are able to carry for considerable time an amount of current largely in excess of their rated capacity, and that seldom, if ever, all of the lamps or the power constituting the installation are utilized, the practice is to install meters of a size somewhat less than the capacity of the installation. Thus in residences the meter capacity is but 50 per cent. of the installation, and in stores 75 per cent. For motors the capacity is based upon an allowance at two amperes at 240 volts per horse-power for commercial motors and three amperes for elevators. Exceptions are made as where the entire installation is devoted to sign and show window lighting, and where it is known that the size of the motor is not greater than the average load on it. But such departures from the general rule are made only after careful investigation at the premises by the Company's local inspector.

PRICES

Current is sold to the public under several schedules, the one selected in each instance being that which offers to the customer the best arrangement that can be made. The price varies only between given classes. One cannot obtain a price better than another's in the class, and to get in any class it is only necessary to comply with the conditions that control it, which are alike to all.

To the smaller customers current is sold under a retail contract in which the maximum price is 15 cents a kilowatt hour, falling as the average use of the installation increases; thus it is 10 cents for the third and fourth hours of average use; 7½ cents for the fifth and sixth hours, and 5 cents a kilowatt hour for all use in excess of six hours daily.

The larger installations are supplied under wholesale contracts, which, through the medium of guarantees, insuring a relatively large consumption of current, offer marked reductions from the prices of the retail schedule. Thus, on what is known as the Wholesale "A" form, a customer guarantees a monthly use of current amounting to 2,000 kilowatt hours, and a daily average use of the installation of two hours, for ten months of the year, and the maximum price is reduced to 10 cents a kilowatt hour. For all current in excess of four hours daily average use of the installation, the price is reduced to 5 cents a kilowatt hour.

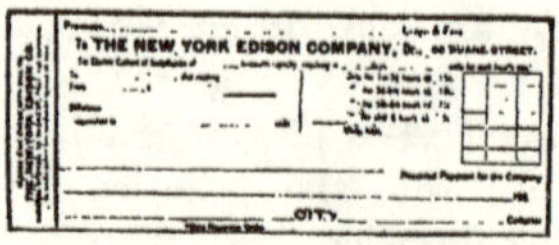

The second wholesale form of contract, known as "B," provides for a guarantee of 2,500 kilowatt hours monthly consumption and two hours daily average use of the installation, for ten months of the year; the maximum price also is 10 cents a kilowatt hour, with discounts of one cent, two cents and three cents respectively for an average daily use of four, six and eight hours; of one-half cent a kilowatt hour where the monthly bill reaches $500, and of one cent where the amount is $1,000. Thus, under this form of contract a customer having a monthly bill of $1,000 and using his installation an average of eight hours daily can purchase current at a price as low as six cents a kilowatt hour.

The large buildings erected on Manhattan Island during recent years have brought even greater problems to the electrical companies, and for these there is a still different schedule which possesses some advantages over the others. Under it the customer guarantees that the monthly use of current shall amount to not less than 10,000 kilowatt hours during twelve months of the year, and places at the disposal of the company, without rental charge, such reasonable space as it may require for rotary apparatus permitting it to use, if desired, high tension feeder service. In view of these considerations the maximum price is reduced to 5 cents a kilowatt hour. Where the monthly use of current exceeds 15,000 kilowatt hours, for the excess alone the price is reduced to 4½ cents; if it exceeds 25,000 kilowatt hours, for the excess above 25,000 the price is 4 cents; if in excess of 35,000 kilowatt hours, the price for this excess is 3½ cents; if in excess of 50,000 kilowatt hours, the price for the excess beyond that point is 3 cents a kilowatt hour. The price remains unchanged in each of the several steps of this schedule and does not include the supply of incandescent lamps or any care of the installation.

Electric current is supplied for power purposes at 10 cents a horse-power hour, equalling thirteen and a fraction cents a kilowatt hour. On monthly bills for 100 horse-power hours there is a discount of 20 per cent.; 200 horse-power hours, 25 per cent.; 400, 30 per cent.; 600, 35 per cent.; 800, 40 per cent.; 1,000, 45 per cent.; 1,500, 50 per cent.; 5,000, 55 per cent., and 10,000, 60 per cent. Thus consuming the largest amount, current may be purchased for 4 cents a horse-power hour.

There is also a special storage battery and automobile schedule for the larger users of that class of service—the smaller users taking current at the power rates. Guaranteeing a monthly use of not less than $50, and agreeing that current shall not be taken during the maximum hours of the district in which the service is given during the months of November, December, January and February, the price is 6 cents a kilowatt hour, with a discount of

½ cent for 3,000 kilowatt hours, of 1 cent for 5,000, 1½ cents for 8,000, 2 cents for 10,000, 2½ cents for 25,000, and of 3 cents for 50,000 kilowatt hours of monthly use. Thus a very large customer in this class of service, evading the maximum hours of the district, can obtain current as low as 3 cents a kilowatt hour.

Service is rendered to private plants only under a special contract which stipulates a monthly use of current amounting to $1.00 annually at the rates of the contract where a reserve is provided for the entire installation, and to $1.50 where but a portion of the installation is provided with reserve. If the consumer will agree to use no current during the maximum hours of the district, these guarantees are reduced one-half, namely, to 50 and 75 cents respectively. Question frequently arises as to the wisdom of thus rendering assistance to private plants, but experience has shown that this is one of the ways to secure them for full central station service. As a medium to that end it has been productive of good results.

No distinction is made between arc and incandescent lighting, both being supplied where desired at the same rates and through the same meter. It is also the practice to supply heating and cooking apparatus under the power or any of the other schedules; if supplied at the retail rates, this apparatus is not included in the rating of the installation. Ventilating motors used during the summer months are also excluded from the rating; thus in increasing the apparent average use and reducing the cost of current is their use encouraged.

INCANDESCENT LAMPS

Experience has shown that where customers who are not informed technically purchase incandescent lamps upon the open market, their standards of lighting diminish materially, causing not only dissatisfaction to the immediate user but a very poor opinion generally on the part of prospective users. For this reason the Company requires that all retail customers shall permit it to supply incandescent lamps, including them within the price for current. The wholesale customers in the private plant class, however, have the option of purchasing incandescent lamps either from the Company or elsewhere, and when this option is exercised a special reduction of one cent a kilowatt hour is made.

HISTORICAL

The Edison system of to-day, as has been briefly outlined, has grown from the original Pearl Street

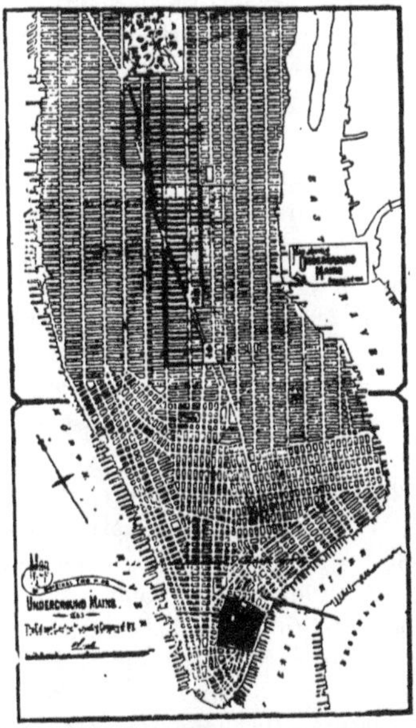

Showing in the lower portion of the map the Edison system in 1883. The mains outlined in the upper portion were laid at a later date

Station, containing when started six steam units, the historical "Jumbos," and supplying current to an underground system of less than fifteen miles in mains and feeders, occupying a territory of which Wall street was the southern boundary and Nassau street the northern—about a mile square.

The Pearl street property—255-257—was purchased in May, 1881. Work of laying street mains and wiring prospective customers' premises began in July the same year. The boilers were started on June 29, 1882; the first engine, installed for operating a coal conveyor, was placed in service on June 30th; the first dynamo was started on July 5th and first generated current on July 8th, supplying 1,000 lamps arranged in a bank on one of the upper floors; the underground system was connected and tested during July, and on Setpember 4, 1882, at 3 o'clock in the afternoon, the station was placed in permanent operation. It ran continuously with but one break of about three hours, in 1883, until the fire of January 2, 1890. Through the assistance of the Liberty street annex, placed in commission in September, 1887, the interruption of the service after the fire did not last more than half a day; thus, since 3 o'clock of the afternoon of September 4, 1882, until the present time, the Edison service on Manhattan Island has been fully interrupted only twice, and the aggregate of these interruptions has been less than twelve hours.

This first district station and distributing system were developed personally by Mr. Edison. For many months, day and night, the work had his constant and direct supervision. That which is to-day best in the generation and distribution of electric current is proceeding on lines which he discovered and followed then; the direct-connected unit, the underground system, the feeder system, safety fuses, the meter, in addition to the high resistance incandescent lamp, the foundation of all. There have been many changes in detail, but basically the best that remains to-day found a place in the original work of Mr. Edison, more than twenty years ago.

IN CONCLUSION

This statement might be properly concluded with reference to the 1883 annual report of the company, which contains a résumé of the work of the preceding fifteen months. Speaking of the results to that time accomplished, it says they consisted of "a great achievement scientifically and commercially, which it is gratifying to recall." Its readers are reminded that in 1875 many of the scientific men of Great Britain before a select committee of the British House of Commons declared unanimously as their best belief the impossibility of the subdivision of electric light; and states that at the very time Mr. Edison had already accomplished this subdivision in a manner which he believed capable of commercial application.

The report continues that the task which Mr. Edison had set for himself was to devise and set in successful operation commercially a system by which electric current could be generated and distributed from a central place to all buildings in a town or other common area, turned on or off by the householder at will without danger or inconvenience, to furnish a light healthful and agreeable to the eye, in quantities suited to domestic habits and pursuits, for a price which the consumer would be able and willing to pay, and which would return a satisfactory profit to the investor.

The aims so clearly set forth in this report, now added—may be looked upon largely, if not entirely, in the light of accomplishment.

NOTE

Electric Cab Service in New York

It has been said, and in the past, rightly, that of all of the cities of the world, the cab service of New York is the most expensive and at the same time the most unsatisfactory. True at least in part in the past, but not so to-day. For a change has been accomplished by the New York Transportation Company. Starting seven years ago with twelve electric vehicles, it is now operating on Manhattan Island a service of five hundred and sixty-five vehicles of various types, including victorias, surries, hansoms, runabouts and tonneaus, many of exclusive design. These vehicles aggregate a daily travel of five thousand five hundred miles—two million miles annually. Curiously in this distance they consume almost exactly two million kilowatt hours, an average of one kilowatt hour a mile of vehicle travel.

The Transportation Company occupies three stations, Eighth avenue and Forty-ninth street, 230-250 East Sixty-sixth street, and the corner of Vesey street and West Broadway. The latter station is for the use of vehicles going downtown to the Bridge and Wall street district. The Sixty-sixth street station contains 30,000 square feet of floor space, the Eighth avenue 128,000 square feet. Exclusive cab privileges are held at the Thirty-fourth street station of the Long Island Railroad, at Sherry's, at the Café des Beaux Arts, and, when opened, at the new Hotel Astor on Long Acre Square.

So many people desire to see New York city in a very short period of time, that the Transportation Company has organized a number of sight-seeing tours, one of which includes a nine-mile ride through Central Park and the Riverside Drive, bordering the Hudson river--said to be the most beautiful residential avenue in the world—past the Soldiers' Monument to the Tomb of General Grant, which overlooks the Hudson river at One Hundred and Twenty-third street. Electric bus service is also maintained on Fifth avenue from the Arch of Washington Square to Seventy-second street.

The schedule of prices is published, and to obviate the possibility of overcharge, the drivers give a duplex form of receipt for all payments which enables the patrons of the company to make any desired complaints, and the company to locate the offender. Thus it would seem that in this modern service the public is fully protected against unjust charges. Cabs cannot be hailed from the street, but must be engaged at one of the stations of the company.

The vehicles manoeuver in and out through the traffic of the streets much easier and faster than would be practicable were they horse-drawn. It is estimated that the speed with which electric transportation is accomplished is about twice as fast as with horses. The vehicles are entirely odorless, easy riding, and the weather, whether extremely hot or extremely cold, makes no difference—there is no horse to be injured by sunstroke or by falling on slippery pavements.

With the growth of this method of transportation New York is no longer compelled to take a second place amongst the cities of the world in the facilities offered in private conveyances to its own citizens and to visitors from abroad.

WESTERN UNION TELEGRAPH COMPANY

Main Building Western Union Telegraph Company, 195 Broadway

Western Union Telegraph Company —Its Origin, Growth and Development

THE history of The Western Union Telegraph Company goes back to the year 1851, when articles of association of The New York and Mississippi Valley Printing Telegraph Company—the original name of the Company—were filed at Albany. During the years that had interevened between the construction of the first telegraph line between Baltimore and Washington in 1844, and the organization of this company, more than fifty different telegraph companies had sprung up in various parts of the United States, largely as the result of quarrels and differences among those interested in the original Morse patents. Most of these companies were licensed by the owners of the patents; a few used the devices of Alexander Bain, and others the printing telegraph system invented by Royal E. House. Lines to operate the House instrument had already been built between New York City and Boston, and between New York City and Philadelphia, and it was for the purpose of extending this system westward that a company was organized in Rochester, N. Y., that afterwards became the Western Union Telegraph Company.

The absurdity of continuing so large a number of independent organizations under the adverse conditions then existing soon became apparent, and as a natural consequence local consolidations began to be effected. One by one, by lease, by purchase, or by exchange of stock, the various companies in the west came into, or were absorbed by the new company, which, by an act of the New York Legislature in 1856 had its name changed from The New York and Mississippi Valley Printing Telegraph Company to the shorter and more popular title of The Western Union Telegraph Company, indicating the union of the western lines into one compact system.

The next important step was taken in 1861, when a line was constructed across the western plains to connect the eastern system of lines with those of the Pacific coast. Meanwhile, as might have been expected, as the result of consolidations, a new era of prosperity dawned upon the telegraph, and with the growth in population the business of The Western Union Telegraph Company rapidly increased. This increase made an outlet to the eastern seaboard cities extremely desirable, to secure which the Company acquired such lines as gave it an entrance into both New York City and Philadelphia with full control of the intervening territory.

From the date of its organization at Rochester to the removal of its offices to New York City in 1866, the Western Union Company had virtually absorbed all rival and opposing companies.

The commanding position reached by The Western Union Company in 1866, with its growing ramifications covered by 75,000 miles of wire, has been maintained until now—the largest telegraph company in world—it embraces in its great system 1,089,212 miles of wire, with 196,517 miles of poles and cables; it operates 23,120 separate offices, two Atlantic cables, two Cuban cables with connections throughout the West Indies, and it has close direct connections with all parts of South America, as well as exclusive connection with the Great North Western Telegraph Company of Canada.

The number of messages transmitted by the Company during the year 1903 was 69,790,866, excluding those sent over leased wires or under railroad contracts, the equivalent of which would probably number 10,000,000 additional messages.

Such an immense volume of business can only be handled successfully by the use of machinery and appliances of the most modern and efficient character. Dynamo machines and storage batteries have, therefore, to a very large extent, been substituted for chemical batteries in recent years, there being now nearly 700 of the former, and upwards of 18,000 of the latter distributed amongst the principal offices, which formerly depended upon the gravity battery to furnish the necessary currents. Copper wire of great conductivity and endurance has been, and is still fast replacing the more perishable iron wire. Large and substantial pole lines have taken the place of the slender and imperfectly insulated pole lines formerly used, whilst the instruments and apparatus employed in the transmission and reception of business are all of the latest and most approved type.

THE MAIN OFFICE

The general office of The Western Union Telegraph Company, at 195 Broadway, is, with one exception, the largest telegraph office in the world. The building in which the office is situated was designed and built to meet the peculiar requirements of the telegraph service. It is thoroughly fireproof in construction, every precaution having been taken in fitting it for use to prevent the possibility of a recurrence of such a disaster as destroyed the old office on July 18, 1890.

Four thousand and two hundred wires enter the building from underground conduits. These wires are led in cables of one hundred conductors each through iron ducts to cable heads in the terminal room of the cellar. Each cable head is equipped with an iron frame containing fifty-two insulating binding posts on opposite sides, and a runway frame for the distribution of wires from one cable to another, the arrangement being such as to greatly facilitate the rapid changes desirable in cases of failure of any conductor in use.

Attached to one portion of the frame work are slate terminal boards, from which connections are made to the 3,800 wires distributed throughout the operating rooms, all wires being equipped with fuses so as to

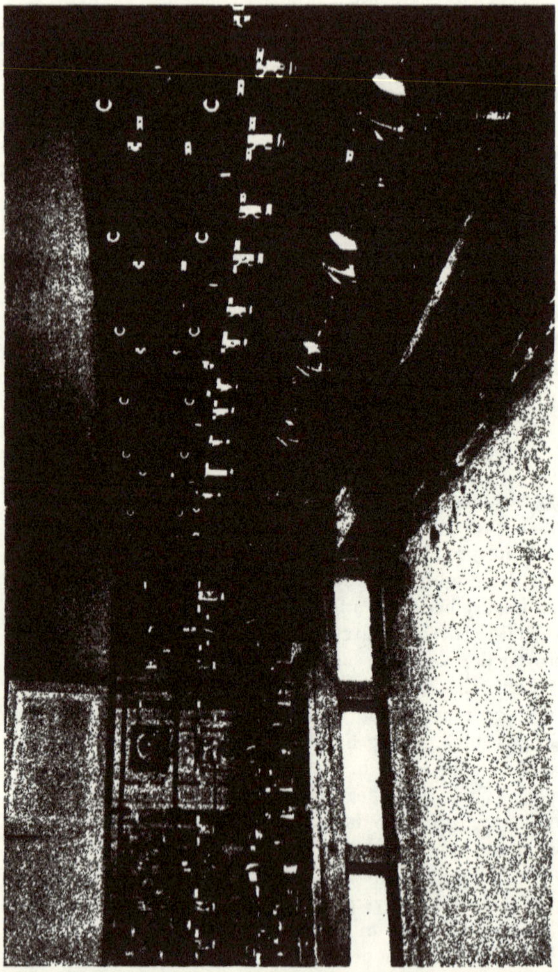

Motor Dynamo Plant

protect the instruments from electric light or other stray currents that might endanger their safety.

In the operating rooms on the seventh and eighth floors are five main-line switchboards of the latest double springjack pattern, separated as widely as possible from each other, and placed as far as practicable from elevator shafts and stairways, an arrangement which not only affords greater security from fire, but permits each section of the switch to be located in that particular division where its wires are operated. In addition to these main-line switches, which have a total capacity for 1,350 wires, there are a number of others used for special purposes, including the loop switch, which is a collection of 676 springjacks with 1,300 cords and wedges. Every quadruplex and duplex loop entering the office is led directly to this board, the arrangement being such that any loop may be connected to any multiplex set, and by means of "flying cords," of which there are 175, any wire from any of the other switches may be brought to this switch and thence to any part of the operating rooms.

The seventh floor contains the Wheatstone automatic and Buckingham printing appliances; the Gold and Stock Quotation and Commercial News departments, and the Filing and Service departments.

About fifteen of the new electrically operated keyboard perforating machines are required for preparing messages for transmission by the Wheatstone instruments, and nearly as many typewriters copy the received slip for delivery. Three Wheatstone duplex circuits are in operation; two to Chicago and one to St. Louis, at speeds of from 130 to 175 words per minute, with one repeater in circuit at Buffalo and Pittsburg, respectively.

Two automatic printing telegraph circuits, each having a capacity of 200 messages per hour, are in operation, one between New York and Chicago, and the other between New York and Buffalo. Messages by this system are transmitted in opposite directions at the rate of one hundred words per minute.

For the Gold and Stock and Commercial News

service, five keyboard machines run by small electric motors are in use for sending out quotations to as many classes of tickers, namely, produce, cotton, petroleum, coffee and financial news. The switchboards devoted to this service are of special form and construction.

A large space on this floor is taken up by the City and Marine departments, into which all the wires from city and suburban points are led through an eight-section springjack switchboard having capacity for 1,000 wires.

On the eighth floor are the wires to the Southern, Eastern and Western points. Here are to be found most of the simplex, duplex and quadruplex circuits; the automatic and quadruplex repeater sets; the time repeater apparatus; the leased wire annunciator system; the loop switch, and other accessories by means of which business may be promptly transmitted and delivered.

On the gallery, in the centre of this floor, the entire business of the two floors is handled by a corps of assorters and distributing clerks. The pneumatic tubes are on this gallery in connection with the receiving and delivery department on the basement floor, and with the several offices and floors of the building, also with downtown exchanges and with the uptown offices. There are also two carrying belts to facilitate the passage of business to and from the penumatic tubes and carriers. Messages between the seventh and eighth floors, and from any section of one floor to other sections of either floor are handled by a cable carrier system. This system consists essentially of a number of double-track miniature cable railways radiating from the central station on this gallery. The carriers are so constructed that they drop the driving cable and are automatically "sidetracked" at the particular stations on the floor for which they are intended, as well as pick up the business at any particular station for which they have been adjusted.

The instruments and apparatus for handling and forwarding messages represent all of the latest improved methods of telegraphic transmission, the most noteworthy examples of which, perhaps, are the alternating current quadruplex system, and the high speed duplex printing telegraph system now in practical operation over several of the company's lines.

Ticker System; Transmitter

Altogether there are in the operating department 350 simplex sets, 92 quadruplex sets, 72 duplex sets, 51 repeaters of various kinds, as well as 637 simplex, duplex and quadruplex loops.

In a room in the cellar of the building are the

dynamos which supply the currents for telegraph work. There are three groups of dynamos supplying main-line current, each consisting of five Edison 40-ampere machines connected in series. The first and second of each series have a potential difference of 70 volts each, and the third, fourth and fifth, 60 volts each. Leads from each machine carry currents of 70, 140, 200, 260, and 320 volts to resistance lamps in the operating department. One group of dynamos furnishes positive and one negative currents. The third is a spare or relief group. There are also two 80-ampere machines supplying currents at 30 volts for loops, two 300-ampere machines furnishing currents at 7 volts for locals, and two 40-ampere machines supplying currents at 45 volts for city and short line service. Of these two lower voltage machines only one of each pair is in use at any one time, the other being held in reserve. There are in this room in addition thirty-five small dynamos for use as intermediate main-line batteries, which deliver currents at from 50 to 125 volts.

The building is lighted throughout by electricity, the current for which is obtained from the mains of the New York Edison Company. Steam for driving the various engines is supplied from the underground street mains of a central steam service; but for security against accidental interruption to this supply, there are in the building six boilers always kept ready to start up at a moment's notice. Electric light sources of supply that can be tapped and utilized in cases of emergency also afford additional security.

The following statistics with regard to the length of cables and conductors in use by the Western Union in New York City may be given:

Kind of Cable.	Total Length of Cables in Feet.	Total Length of Conductors in Feet.
Aerial	1,337,160	21,941,874
Underground ..	314,931	12,766,986
Submarine	164,378	2,684,048

The volume of business handled in the main office varies slightly at different seasons, but the daily average for the year is over 100,000 messages, exclusive of press reports.

THE NEW YORK CENTRAL CABLE OFFICE

Although a branch office in character, the New York Central Cable Office is one of the greatest and most important telegraphic centres in this country, for here is concentrated the great bulk of the telegraph business of the financial district of the metropolis of America.

This office occupies the ground and basement floors of the south end of the magnificent new Stock Exchange Building, the area included being 40 by 160 feet, extending through from Broad to New streets. The operating room, which is a marvel of completeness in modern telegraphic equipment, is on the ground floor at the Broad Street end, while at the rear is found the office of the American District Telegraph Company and of the Gold and Stock Telegraph Company.

In order to handle this telegraphic traffic, eighty-five aerial lines and two submarine cables centre here, while seventy-four operators are employed besides a clerical force of fifty-three.

The longest circuits maintained continuously anywhere in the world are operated from this office. One of these is a duplex to San Francisco, upwards of 3,000 miles in length, over which the traffic to the Pacific coast passes. There are four duplexes leading to Galveston, Texas, 2,200 miles away, where connection is made with the Mexican and South American cable systems. Two direct duplexes extend to Havana, 1,800 miles distant, there connecting with the Cuba Submarine Cable Company's system, and the land lines of the Insular Government. Five duplexes reach out to North Sydney, C. B., and are employed to carry the traffic of the European cables landing in its vicinity. A duplex to Duxbury, Mass., connects with the cable landing at that place. Every important centre in the United States is in direct communication with this office. To Chicago, two duplexes are operated; two to New Orleans; three to Boston; several connect with Philadelphia, and others

Section of Operating Room

bring St. Louis, Cincinnati, Baltimore, Toronto, Montreal and other places into close touch with the financial pulse of the Western Hemisphere.

The office is connected with the floor of the Stock Exchange in the same building, by an extensive pneumatic tube service. The Stock Exchange floor is divided into eight sections, and sixteen tubes are in constant use to carry from and to each of these sections the telegraph business of the exchange. An average of but nineteen seconds is required in the transfer of such messages. Similar tube connection is also maintained with the operating room at the main office, No. 195 Broadway.

The American District Telegraph Company, which has the largest plant of its kind in the world, has in operation twenty-six call circuits with a total number of 1903 call boxes, all within a radius of a quarter of a mile from this particular office. From 1,000 to 1,500 calls are answered daily, the number being regulated by the activity of the stock market. The number of boys employed will average one hundred, although frequently two hundred are brought into requisition.

The Gold and Stock Telegraph Company, incorporated in 1867, now divides the entire Stock Exchange ticker reporting business with the New York Quotation Company. By special agreement with the New York Stock Exchange, both companies receive their quotations at the same moment, by the same wire.

The New York Quotation Company furnishes the quotations by ticker to members of the exchange only, and the Gold and Stock Company by ticker to all other business houses whose applications may be approved by the Stock Exchange. Its ticker circuits embrace New York City, Brooklyn, Jersey City, Newark, Elizabeth, Patterson and other nearby points, about 670 stock tickers being in operation. This Company also receives and issue by ticker, quotations and market reports from the Chicago Board of Trade, the New York Produce Exchange, and the Cotton Exchange, as well as general markets and prices of petroleum and coffee. On these systems it has over three hundred tickers.

Automatic Repeater Equipment

The systems are operated by a large storage battery plant and central transmitting machinery. It has over 1,000 miles of cable conductors in use in the city, as well as some outlying air lines wires on Western Union pole lines. A force of about sixty men is employed in attending to the service and in keeping the cables and wires in good order.

There are two submarine cable circuits in operation between this office and Canso, Nova Scotia, from which latter point direct communication with the cable station in Penzance, England, is maintained. Both circuits are worked duplex by the automatic system at rates of speed varying from twenty-five to thirty-two words per minute.

The terminal room in the basement, which is the second largest in the Western Union service, has accommodation for four thousand wires. Thirty-one hundred conductor cables enter this room, which is the general distributing point to all the lower part of New York City.

A single Crocker-Wheeler motor-dynamo supplies the current for all short wires to brokers and other outside offices, as well as to all the local sounders in the operating room.

The following tables show the organization and present officers of The Western Union Telegraph Company:

Robert C. Clowry, President and General Manager.
George J. Gould, J. B. Van Every, Thomas F. Clark, Vice-Presidents.
A. R. Brewer, Secretary.
M. T. Wilbur, Treasurer.
J. B. Van Every, Auditor.
John F. Dillon, General Counsel.
G. H. Fearons, General Attorney.
Rush Taggart, H. D. Estabrook, Solicitors.

BOARD OF DIRECTORS

Thos. T. Eckert, Chairman.

Robert C. Clowry.	Henry M. Flagler,
John T. Terry,	John Jacob Astor,
Russell Sage,	Oliver Ames,
Samuel Sloan,	C. Sidney Shepard,
George J. Gould,	John B. Van Every,
Edwin Gould,	James Stillman,
Louis Fitzgerald,	Thomas F. Clark,
Jacob H. Schiff,	W. Lanman Bull,
James H. Hyde,	Morris K. Jesup,
Frank Jay Gould,	E. H. Harriman,
Charles Lanier,	Charles Lockhart,
J. Pierpont Morgan,	Samuel Spencer,
Chauncey M. Depew,	Howard Gould,

John J. Mitchell.

EXECUTIVE COMMITTEE

Thos. T. Eckert, Chairman.

Robert C. Clowry,	Edwin Gould,
John T. Terry,	Louis Fitzgerald,
Russell Sage,	Jacob H. Schiff,
Samuel Sloan,	James H. Hyde,
George J. Gould,	Frank J. Gould.

J. C. Barclay, Assistant General Manager, New York.

C. H. Bristol, General Superintendent Construction, New York.

William Holmes, Superintendent of Tariff and Check Bureaus, New York.

G. W. E. Atkins, Superintendent of Contract and Free Service Departments, New York.

E. C. Cockey, Superintendent of Supplies, New York.

H. E. Roberts, Assistant Superintendent of Supplies, New York.

W. J. Dealy, Superintendent of Commercial News Department, New York.

S. C. Mason, Storekeeper, Chicago.

Eastern Division General Superintendent—B. Brooks, New York.

Eastern Division District Superintendents—

1. E. M. Mulford, New York.
2. E. P. Griffith, New York.
3. E. B. Saylor, Pittsburg, Pa.
4. J. P. Altberger, Philadelphia, Pa.
5. C. F. Ames, Boston, Mass.
6. D. C. Dawson, St. John, N. B.

Western Division General Superintendent—Theodore P. Cook, Chicago, Ill.

Western Division District Superintendents—

1. F. H. Tubbs, Chicago, Ill.
2. G. J. Frankel, St. Louis, Mo.
3. C. B. Horton, Omaha, Neb.
5. C. Corbett, Cleveland, O.
6. John F. Wallick, Indianapolis, Ind.
7. I. N. Miller, Cincinnati, Ohio.
8. James Swan, Minneapolis, Minn.

Southern Division General Superintendent—J. Levin, Atlanta, Ga.

Southern Division District Superintendents—

1. F. E. Clary, Richmond, Va.
2. J. M. Stephens, Atlanta, Ga.
3. B. F. Dillon, Jacksonville, Fla.
4. J. R. Terhune, Nashville, Tenn.

Pacific Division General Superintendent—Frank Jaynes, San Francisco, Cal.

Pacific Division District Superintendents—

1. F. H. Lamb, San Francisco, Cal.
2. T. W. Goulding, Seattle, Wash.

European Agency—D. Le Rougetel, General Superintendent, London.

Havana Agency—Eugenio Fortun y Varona, Manager.

NEW YORK CENTRAL & HUDSON RIVER RAILROAD COMPANY

Description of the System Adopted by the New York Central & Hudson River R. R. Company for Operating its Passenger Traffic by Electricity in New York City and Vicinity

FOR the past six years the New York Central and Hudson River Railroad Company has given serious consideration to the question of using electricity as a motive power for handling its passenger traffic from the Harlem River to the Grand Central Station in the City of New York. The necessity for this investigation arose from the serious inconvenience of the traveling public due to the use of steam locomotives in the four-track Park Avenue Tunnel, which extends from 56th Street to 96th Street, a distance of about two miles. In addition to relieving the discomfort to passengers, the use of electricity was deemed advisable for minimizing the annoyance from noise, gas, steam and cinders to abutting property owners along Park Avenue.

Furthermore, the City of New York desired the lowering of the tracks of the Grand Central yard and terminal from 56th Street south to 42nd Street, so as to permit the restoration of the cross streets from 45th Street to 56th Street, inclusive, which heretofore have been cut in two by the approaches to the terminal. This depression of the yard and terminus, with the consequent roofing in of the tracks by streets and viaducts was not feasible with steam locomotive operation, and hence arose an additional necessity for using electricity as a motive power.

Plan of one of the Armatures

The natural conservatism of a large corporation like the N. Y. C. & H. R. R. R. led to a more than ordinarily careful and deliberate consideration of such a revolutionary step as the change of motive power, from the long and well-tried steam locomotive to a new force which had not yet been employed for the peculiar character of service existing on one of the principal trunk lines of the country. The hauling of trains nearly one-quarter of a mile in length and weighing as high as 871 tons, at speeds exceeding sixty miles per hour, with regularity and safety, had not been attempted by electricity. The handling of from 500 to 700 trains a day in the Grand Central yard, with its maze of tracks and switches, presented a problem the solution of which has not before been attempted with electricity as a motive power. Moreover, the difficulties presented to the N. Y. C. & H. R. R. R. were not like those which arise in the majority of electric railroad enterprises, where the change of power is not attended by the embarrassment, delays and dangers incident to the successful maintenance of an existing enormous traffic in conjunction with radical changes in roadbed and structures.

After a thorough investigation of the entire subject the Company finally decided, in 1902, to change its motive power as well as make other radical improvements, and in the following year the State and municipal authorities sanctioned the carrying out of these alterations.

In addition to the change from steam to electric power, it is the intention of the Company to radically improve and enlarge the Grand Central Station, increasing the size and depressing the yard in such a manner as to permit the restoration of cross streets from 45th Street to 56th Street, inclusive; build a new underground suburban terminal connecting with the Rapid Transit Subway beneath the Grand Central Station; four-track both the Hudson and Harlem Divisions within the electrical zone; eliminate all grade street and track crossings within the same zone; straighten the alignment in the Borough of the Bronx by what is designated as the Marble Hill Cut-off; and make important station improvements at places like Yonkers, Ossining, Mount Vernon and White Plains.

Electrical engineers will, of course, be most interested in the electrification of traffic, and so far as possible this description is confined to that part of the work.

The Act of the Legislature under which these improvements are authorized calls for the completion of the change of motive power from the Harlem River to the Grand Central Station within five years from the date upon which the law went into effect, July 1, 1903. The Railroad Company decided that for many reasons it would not be advisable to change from electric to steam locomotives at the Harlem River, which is about six miles from the Grand Central Station, but to extend the electrical zone to a point on the Hudson Division between Ossining and Croton, a distance of about thirty-four miles, and to a point near White Plains on the Harlem Division, a distance of about twenty-four miles from the Grand Central Station, involving the electrification of nearly 300 miles of single track and carrying the most important passenger and mail traffic in the country.

The Electric Traction Commission, under whose charge the change of motive power was placed, arrived at the conclusion that the electrical zone should be extended to points considerably beyond the limit fixed by the Act of the Legislature.

The passenger traffic was found to divide itself naturally into two categories, namely, suburban service and through train service.

In considering the suburban service, which now extends to Croton and White Plains, the Electric Traction Commission felt that through the use of smaller trains and by a more frequent and faster service, with freedom from smoke and cinders, a large increase in traffic would follow. The members of the Commission felt that these advantages, with the resultant increase of population within the suburban zone, could only be obtained by uniformity of motive power within its limits. The topographical restrictions preclude adequate terminals at any point near the Harlem River, or for that matter at any point south of the present limits of the suburban traffic at Croton and White Plains. Furthermore, the vacant territory in the northern portion of the City of New York is growing so fast that they believed it unwise and impolitic to establish new steam engine terminals within the city limits. These considerations, among others, led the Company to adopt the plan of an electrically operated zone south of Croton and White Plains.

After deciding upon the limits of the electrically operated service, the next step was the selection of the character of current to be used. The Commission, for several reasons, unanimously adopted direct current. Among these were that direct current would facilitate future interchange of equipment with other rapid transit lines in New York City and its vicinity already similarly equipped. A number of legal and physical conditions absolutely prohibited the use of overhead conductors for alternating current on the Park Avenue viaduct and in the Park Avenue tunnel. Alternating current apparatus had not been developed to the point where it was considered practicable for such an important installation which must be operated successfully on a fixed date. Therefore, disregarding all debatable questions of relative costs of operation and apparatus, there remained strong reasons of policy and certain technical advantages in favor of the adoption of the direct current system within the proposed electrical zone. North of Croton and White Plains there still remains a full opportunity for the future adoption of the alternating current system when its success has been demonstrated for long-distance service.

As a result of the Commission's conclusions the Railroad Company has placed orders for the larger part of the new equipment, including electric locomotives, power stations, turbo-generators, surface condensers, boilers,

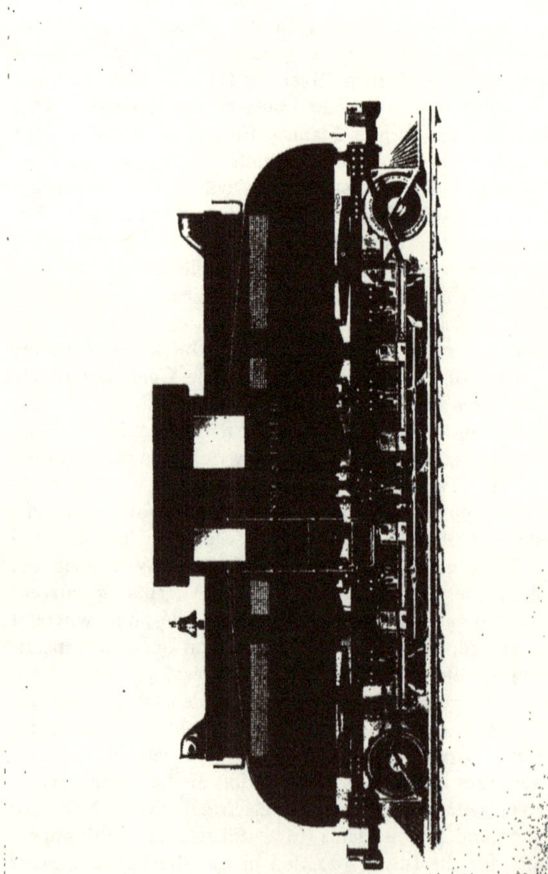

View of New York Central Electric Locomotive

feed-pumps, exciters, feed water heaters, and other auxiliary power station apparatus. The actual placing of these orders marks a profoundly important epoch in the application of electricity to the hauling of trains on the main trunk line of one of the leading railroads of the world. For the first time electric motors will compete in performance with steam in hauling heavy express trains of 500 tons and over at speeds exceeding sixty miles per hour. The change is expected to show marked economies in the electrified part of the railroad as a whole, and it will mean a great increase in the suburban traffic to Croton and White Plains, due to the purification of the tunnel and the running of multiple unit trains every few minutes.

To ensure the utmost reliability of the service, two central power stations are to be erected, one at Port Morris and the other in the vicinity of Yonkers, each of an ultimate capacity of 30,000 kilowatts, and so connected that either is able in case of accident to carry the entire load of a train service much greater than the present steam schedule.

After an exhaustive examination of the relative merits of reciprocating engine driven alternators and turbo-generators, the Commission has recommended the use of the latter, and the contract has been executed for eight 7,500 H.P. turbo-generators with an ultimate installation of twelve. The turbines are the Curtis four-stage vertical type running at 500 revolutions. The generators are twenty-five cycle, three-phase, alternating current, wound for 11,000 volts. The decision of the Commission in adopting turbo-generators is based upon a considerable saving in first cost, space occupied and guaranteed sustained steam consumption. The turbo-generators, while normally rated at 7,500 H.P., can readily develop over 10,000 H.P.

The condensing apparatus will be external to the turbines. The condensers are of the counter current surface type and each is directly connected to its turbine base and contains 17,000 square feet of tube surface. The condensing apparatus is guaranteed to maintain a vacuum of twenty-eight inches, with cooling water at a temperature of 70° Fahrenheit. The auxiliary condensing machinery is composed of independent units. Circulating water pumps are of the centrifugal type, directly driven by horizontal reciprocating engines. The dry vacuum pumps are of the rotative flywheel type, with the air and steam cylinders in tandem erected on a common base. The hot well pumps are of the centrifugal type and are driven by direct connected electric motors. As an evidence of the high efficiency expected from the condenser system, it may be stated that the manufacturer has guaranteed that the temperature of condensed steam measured in the condenser hot well will be within one degree Fahrenheit of that corresponding to the pressure measured in the condenser. All parts of the machinery have been designed to operate smoothly and quietly under all loads up to fifty per cent. above the normal rated capacity of the turbines.

The exciter equipment in each power station will consist of two turbine driven steam exciters and one induction motor exciter. Each unit is of sufficient capacity to operate the entire station. One spare exciter unit will be used for operating the electric traveling crane and station lighting. In addition to these there will be a storage battery exciter located on the top floor of each switch house, having a capacity of 1,200 amperes for three hours.

The turbine room of each power station will be equipped with a fifty-ton electric traveling crane having an auxiliary ten-ton hoist. The cranes are of the usual power station type.

The boiler houses will be of the one-floor type, and will contain eventually forty-eight 625 horse-power water tube boilers with internal superheaters. The boiler headers are of forged steel and the tubes are staggered in serpentine rows. There are twenty-one sections and fourteen tubes in each section. The total heating surface of each boiler is 6,250 square feet, and the grate surface is eighty-eight square feet. The boilers are designed for a normal working pressure of 185 pounds, and the steam will be superheated to 200° Fahrenheit over and above the temperature due to the steam pressure. The superheaters in each boiler contain 1,230 square feet and are made up of 168 two-inch "U" shaped tubes each 13′ 5″ in length. The manufacturer has guaranteed that the economy of each boiler and superheater at the normal rated capacity, under the usual working conditions shall be not less than the equivalent evaporation of ten and three-fourth pounds of water from and at 212° Fahrenheit per one pound dry, run-of-mine, bituminous coal, containing approximately 14,000 British thermal units.

Each section of the power station containing four boilers is equipped with one boiler feed pump. The boiler feed pumps are of the duplex, outside packed piston type. The pumps are designed for hot water and each pump has a capacity sufficient to feed eight boilers under full load conditions.

The adopted feed water heaters are of the closed type, this decision having been reached after a very careful consideration of the quality of the feed water, and due consideration given to the fact that the water in the hot well from the surface condenser plant would be absolutely free from oil. The heaters are of the copper corrugated tube, Wainwright, counter current design.

There will be two chimneys for each power station, constructed of radial brick, carried directly over the boilers on steel piers. The decision in favor of radial brick stacks was reached after a consideration of the relative merits of radial brick stacks and steel stacks, and resulted in the adoption of radial brick stacks on account of their permanency and relatively lower cost as compared with lined, self-sustained, steel plate stacks.

The high-tension switching equipment of each power station is to be located in a separate switch house, the high tension leads from the generators running direct without break to the oil switches. In the gallery of each power station is located the operating or pilot switchboard and there is in addition an emergency switchboard in each switch house.

The system of electrical distribution will be 11,000 volt, three-phase, alternating current, generated direct and fed to eight substations, where the primary current will be transformed into 600-volt continuous current and fed directly into the working conductors. The working conductors will consist of the usual third rail, except at crossings, and in complicated yard work, where an overhead rail will be installed, the motors automatically taking current either from the third rail or the overhead conductor. The third rail will be provided with a protecting shield against accidental contact and from sleet. The third rail will be augmented by an auxiliary feeder running throughout the entire system to ensure the continuity of service should certain sections be cut out through accidental or other causes.

The substations will be supplied with double sets of conductors and the power houses will be connected together, so that all substations can be operated from either of the main generating stations.

For the system as now decided upon, there will be the eight substations, each equipped with rotary converters varying in capacity from 1,000 to 1,500 kilowatts direct current output. The advisability of installing electric storage batteries to "float on the line" is under serious consideration. The substations are laid out with bus bars and high tension connections in the basement; the first floor contains the transformers, rotaries and oil switches. If storage batteries are installed, they will be located over a waterproof floor above the rotaries.

The distributing system will be partly aerial line construction, the conductors being carried on steel poles, and partly in subways, the duct system having manholes either in duplicate or constructed with isolated compartments and separate manhole covers.

All suburban trains will be operated on the multiple unit system, in which trains are made up of a number of motor cars, under a common control, on the general plan now adopted by the Elevated and Underground Railways.

The hauling of heavy through trains presented for consideration problems of more than ordinary interest. The trains vary in weight from 200 tons to over 900 tons, and are operated at speeds of from six miles per hour in the yards to seventy-five miles per hour on the main line. This necessitates at present a great variety, both in style and capacity, of steam locomotives. The adoption of electricity afforded a solution denied to steam in operating this class of traffic, because that with electricity it is possible to adopt, for all classes of service and for all speeds required, a single type and size of electric locomotives, capable of being governed and grouped to meet all exigencies of service. From a variety of plans submitted, the decision was finally made in favor of an electric locomotive having a new type of gearless, direct current, bi-polar motor, and a contract has been awarded for furnishing an equipment of thirty such electric locomotives, with the option of increasing the number to fifty. The total weight of the locomotive is eighty-five tons, of which sixty-seven tons is borne on the four pairs of drivers. Compared with existing steam practice it will be interesting to note that the heaviest Atlantic type locomotive of this company weighs forty-seven tons on the two pairs of drivers, and the total weight, including the tender, is 150 tons; thus for every pound of effective drawbar pull, the steam locomotive weighs twelve and one-fifth pounds, as compared with five and one-fifth pounds for the electric locomotive. This gives in a single motor unit, over thirty-five per cent. greater weight available for traction than the largest steam locomotive now in use in this service, with forty-three per cent. less dead weight and with twenty-nine per cent. less weight on each axle and an entire absence of counter balancing of drivers and twist from the reciprocal action so destructive to the track and roadbed.

The locomotive is the result of several months'

cöoperative designing between the General Electric Company and the American Locomotive Company, the latter company building the frames and the running gear. These electric locomotives represent in a marked manner the latest development of the best experience in the electrical and locomotive fields, and ensure the highest attainment of the art, in both electrical and mechanical design.

The length will be thirty-seven feet over all. The wheel base will consist of four pairs of motor wheels and two pairs of pony truck wheels, the length of the total wheel base being twenty-seven feet; and of the rigid wheel base, consisting of the four pairs of motor wheels, thirteen feet. The diameter of the driving wheels will be forty-four inches, and of the truck wheels thirty-six inches. The driving axles will be eight and one-half inches in diameter.

The frame will be of cast steel, the side and end frames being bolted together at machined surfaces and stiffened by cast steel cross transoms. The journal boxes and axles will be designed to permit sufficient lateral play to enable the locomotive to pass easily around curves of 230 feet radius.

The superstructure of the locomotive is to be of steeple form, so designed as to offer the least practicable wind resistance, consistent with the adequate housing of the apparatus and its convenient operation. The cab is designed so as to afford a better view of the track and signals than is now possible with that of the steam locomotive. The whole of the superstructure is to be of sheet steel with angle iron framing, and the doors and windows of the cab are to be fireproof.

The driving power of the locomotive will be furnished by four 600-volt direct current gearless motors, each of 550 horse-power. This will make the normal rating of the locomotive 2,200 horse-power, with a maximum rating of about 2,800 horse-power, or about fifty per cent. greater than that of the largest steam passenger locomotives now in service.

The armatures will be mounted directly on the axles

and will be centered between the poles by the journal boxes, sliding within finished ways in the side frames. The armature core will be of the iron-clad type, the laminations being assembled on a quill which will be pressed on the axle. The winding will be of the series drum-barrel type. The conductors will be designed so as to avoid eddy currents, and will be soldered directly into the commutator segments.

The commutator will be supported on the quill. The commutator segments will be made of the best hard-drawn copper and will have the ears integral with themselves. The brush-holders will be made of cast bronze and mounted on insulated supports attached to the spring saddle over the journal, maintaining a fixed position of the brush-holders in relation to the commutator.

Unlike the ordinary four-pole motor where the magnetic circuit is made through a separate box casting, the magnetic circuits in this type of electric locomotive are completed through the side and end frames. The pole pieces are cast in the end frames, and there are also double pole pieces between the armature carried by bars which act as part of the magnetic circuit.

The pole pieces will be shaped so that the armature is free to move between them with ample clearance on the sides. As the poles move up and down with the riding of the frame on the springs, they will always clear the armature, and provision is made so that the armature will not strike the pole pieces even if the springs are broken. The field coils will be wound on metal spools bolted to the pole pieces, and will consist of flat copper ribbon.

Proper distribution and division of the weight among axles will be accomplished by swinging the main frames from a system of elliptical springs and equalizing levers of forged steel, the whole being so arranged as to cross equalize the load and furnish three points of support.

The locomotive will be provided with all the usual accessories of a steam locomotive, including an electric air compressor to furnish air for the brakes; it will have whistles, a bell and an electro-pneumatic sanding device and electric headlight at each end. The interior of the cab will also be heated by electric coils.

To control these giant locomotives, and to secure command of the amount of power necessary for all train operations, the Sprague-General Electric system of multiple unit control has been adopted, which enables two or more locomotives to be grouped together and operated as a single unit from either end. By this means, if necessary, four or even six thousand horse-power, is available for a single train, with the motive power under single, and consequently safer, control, out-ranking any possible steam locomotive combination. As a result of the adoption of a single type of locomotive, in place of a dozen or more types and sizes now used, and the method of multiple-unit control, the electric locomotives are available for the extremes of varied service. They can be used equally well for switching, for the lower speed trains, or in combination, for the very heaviest and fastest trains, the variations of speed and power being effected by changing the relation of the motors from four in series to four in multiple, with intermediate steps, and grouping of the locomotive units as may be desired. The locomotives are double-headed, controllable at either end, and, therefore, no turning is necessary. The first of these locomotives is to be delivered within a few months, and comprehensive high-speed tests will be made.

The Railroad Company has now under construction between Schenectady and Hoffmans six miles of experimental track, upon which all electrical road equipment will be thoroughly tested before being delivered for actual service. This experimental track will be built as a model of the construction intended in the New York district, and all appliances which are intended for use in the permanent system will first have a thorough trial at this experimental station.

Although the law under which the change of motive

power is to be carried out allows five years for its completion in the territory designated by that law, it is the intention of the Railroad Company, and its contracts are so drawn, that there is every reason to believe that it will complete the change from steam to electric operation over a much greater territory than that the law requires a year earlier, and it is intended that all the other terminal changes shall be completed at about the same time.

The Railroad Company not only anticipates a marked increase in the comfort and safety of its passengers by the new operating system, but also a decided increase in the suburban business, and what is of great importance, the adoption of electricity makes it possible to reclaim all the overhead space in the territory occupied by its terminal yard in the heart of the city, which, instead of being left open, as obligatory with steam operation, can be utilized for superimposed structures by the company in any manner best suited to its purposes. A much higher speed and a more frequent suburban train service will characterize the suburban service, in which the locomotive is entirely eliminated, and its places taken by motor cars, operated in train combinations by the multiple unit system.

The substitution of electric power will largely reduce the number of locomotive movements at present required in the terminal yards, and in general operation will reduce the cost of maintenance of rolling stock, wear and tear of the tracks and eliminate the destructive effect of corrosive gases from the steam locomotive.

This transformation now being carried into execution by the New York Central Railroad Company, involving, as it does, a change in motive power in addition to terminal changes of extensive character; and maintaining meanwhile the operation of from six hundred to seven hundred train movements per day, makes it the most extensive and important electric traction development now in process of execution.

The organization under which the electrification is being planned and executed is believed to have special merit. The principle and general policy are fixed by an Electrical Commission which holds weekly meetings, and discusses and takes formal action upon the various problems that are presented. When radical differences of opinion develop in the Commission as to the proper course to pursue, outside consulting engineers are called in for consultation. The electrical engineer of the Company acts as secretary of the Commission and thus understands the reasons that guide the Commission in its action.

The detail plans and specifications are prepared by a corps of electrical and mechanical engineers under the Electrical Engineer and have the final approval of the Commission.

The Commission consists of Mr. Wm. J. Wilgus, Chairman, who is Fifth Vice-President of the Railroad Company in general charge of all construction; Mr. John F. Deems, General Superintendent Motive Power; Mr. Bion J. Arnold, Consulting Engineer and President of the American Institute of Electrical Engineers; Mr. Frank J. Sprague, Consulting Engineer and Past President of the American Institute of Electrical Engineers, and Mr. George Gibbs, Consulting Engineer, who also acts in a similar capacity for the Rapid Transit Subway. Mr. Edwin B. Katte is the Electrical Engineer of the Company.

THE TELEPHONE SYSTEM OF NEW YORK CITY

The Telephone System of New York City

ALTHOUGH scarce thirty years have passed since Professor Bell worked out the first practical application of the transmission of speech, the use of the telephone to-day in the United States is so universal that it is recognized as being a factor of the highest importance in the business and social development of the country.

In 1878, the first telephone exchange to be built in the United States was constructed in New Haven. When it was demonstrated by practical operation that telephony was a success, the American public, ever eager to adopt a useful idea, saw the marvelous utility of this invention, and it was not many years before there was a substantial demand for telephone service.

The problem of serving a few people in a small town was soon solved, but to render an efficient service to a large community, to establish an intercommunicating system with thousands of stations, to bring the service to the high point of speed and efficiency demanded by the nervous activity of the large cities, has only been accomplished by the expenditure of a vast amount of money in experimental work, and by bringing to bear on the problem the best minds and the highest trained skill available.

In New York, the completion of a great telephone system demanded by the dense population within a radius of fifty miles from the centre of the metropolis, is apparently only a matter of orderly growth, from year to year.

The telephone problem of New York is an extremely difficult one. Within a fifty-mile radius of the City Hall, there are 200 cities, towns and villages, varying in population from New York, with nearly four millions, to the small suburban villages, with 1,000. The total population of this metropolitan district is at present estimated at a little under six millions. By 1920, the population, by the most careful estimates, is expected to be about ten and one-half millions.

On the theory that the possible limits of a town are bounded by the accessibility of the business centre to the people who work there, it can readily be seen that the development of the telephone, as well as the transportation system, is an important factor in the city's growth. By the last Federal census, the population of the island of Manhattan alone was increasing at the rate of 50,000 per year, and the population of the adjacent outlying regions at an even larger ratio. Although improved transportation has done a great deal to bring about this increase, yet without question the telephone has been a considerable factor in creating conditions favorable to this development.

But this same wonderful city growth and the fact of the constant increase of the telephone habit, doubles the complexity of the problem. It means that not only shall the great demand of the present day be adequately met, but that the telephone plant itself shall be comprehensively planned to provide for the increasing traffic. Add to this the fact that telephone equipment is constantly being worn out by its frequent use and must be replaced, and also that it has been and still is being constantly improved, so that periodically important parts of the plant must be rebuilt to bring it to its utmost efficiency, and, furthermore, that this must all be done without disturbing the service, and an idea may be gained in a general way of the stupendous nature of the New York telephone problem.

OPERATING COMPANIES

The Bell telephone business of this district is in the hands of two companies.

First: The New York Telephone Company, operating in the old City of New York, consisting of the

Boroughs of Manhattan and The Bronx; the county of Westchester, a region containing 450 square miles, extending from the Hudson River to Long Island Sound, with a population of 203,000, and containing such cities as Yonkers, White Plains, Mount Vernon

General Offices of New York and New Jersey Telephone Company, No. 81 Willoughby Street, Brooklyn, N. Y.

and New Rochelle—Yonkers, the largest city, having a population of about 56,000; a small part of Connecticut which comes within the prescribed limit; and part of Rockland County in the State of New York, on the west bank of the Hudson River.

Second: The New York & New Jersey Telephone Company, operating in all of Long Island and Staten Island, and that portion of New Jersey lying within a radius of fifty miles of New York City Hall. The New York & New Jersey Company therefore serves Brooklyn, Staten Island, Newark, Jersey City, Hoboken, Elizabeth, Paterson, the Oranges, and many other important suburban points. The Rockaway resorts on Long Lsland, and the New Jersey resorts in Monmouth County, such as Asbury Park, Long Branch, Ocean Grove, Seabright, Lakewood and Spring Lake, also come within this territory. This company is organized on the same general lines as the New York Telephone Company, making due allowance for the different character of the territory. Its aim is to extend the telephone to every town, village, hamlet, and, indeed, every farmhouse within the territory, and the measure of its success is shown by the fact that it now has installed 75,000 stations, and is adding to this number at the rate of from 12,000 to 15,000 every year. Notwithstanding this tremendous growth and development, the standard of efficiency is maintained at the highest point, and every effort is made to bring each subscriber in touch with every other subscriber, near and remote. This involves a complete system of toll lines ramifying in all directions, which are centered at various points under plans carefully considered by all of the departments concerned.

A third company is also prominently concerned in the telephone affairs of New York. It is popularly known as the "Long Distance Company," which is the operating branch of the American Telephone & Telegraph Company. The wires of the Long Distance Company have their principal centre at New York and are operated in conjunction with the stations and lines of the two companies above-mentioned and are used to connect these with the stations of the principal telephone companies east of the Mississippi, and even beyond that river, connections being made as far west as Omaha.

THE NEW YORK TELEPHONE COMPANY

A description of one of the companies concerned—the New York Telephone Company—will serve as an indication of the methods employed by all. This system not only provides for instantaneous inter-communication between all stations in the great metropolitan district, but also for connection between any one of these stations and any station on the continent reached by the Long Distance wires.

ORGANIZATION

The organization of the New York Telephone Company aims at the highest degree of efficiency, and it has been developed and perfected until at the present time it is regarded as a model by students of organization and economics, and also by telephone experts from all parts of the world. The members of its staff have been carefully selected for their ability and knowledge, and, as promotion comes from within the ranks and as a most liberal policy is maintained in all its relations with its employees, a high degree of loyalty to the interests of the Company is maintained.

As shown in the chart, the president is the executive head of the system. His immediate subordinates are the 1st vice president and general manager; the 2d vice president and secretary; the treasurer, and the auditor. The 2d vice president and secretary is the corporation attorney, and looks after the company's legal interests. The auditor is the responsible bookkeeper and the treasurer the fiscal agent.

The 1st vice president and general manager is responsible for the entire working system, and for the proper conduct of the business he has formed four general departments—engineering, supply, contract and operating.

The Engineering Department, as indicated by its name, is responsible for all the engineering features of the work. All general and detailed plans for the construction of the plant are worked out by this department and all types of equipment, and, in fact, everything which enters into construction, must first be approved and standardized by the chief engineer. A particularly important work done by this department is the so-called advance study. By estimating from past and present conditions, comprehensive plans are made from different standpoints of what the demands on the telephone serv-

Typical Telephone Building, 18th Street and Irving Place, New York

ORGANIZATION
OF THE
NEW YORK TELEPHONE COMPANY

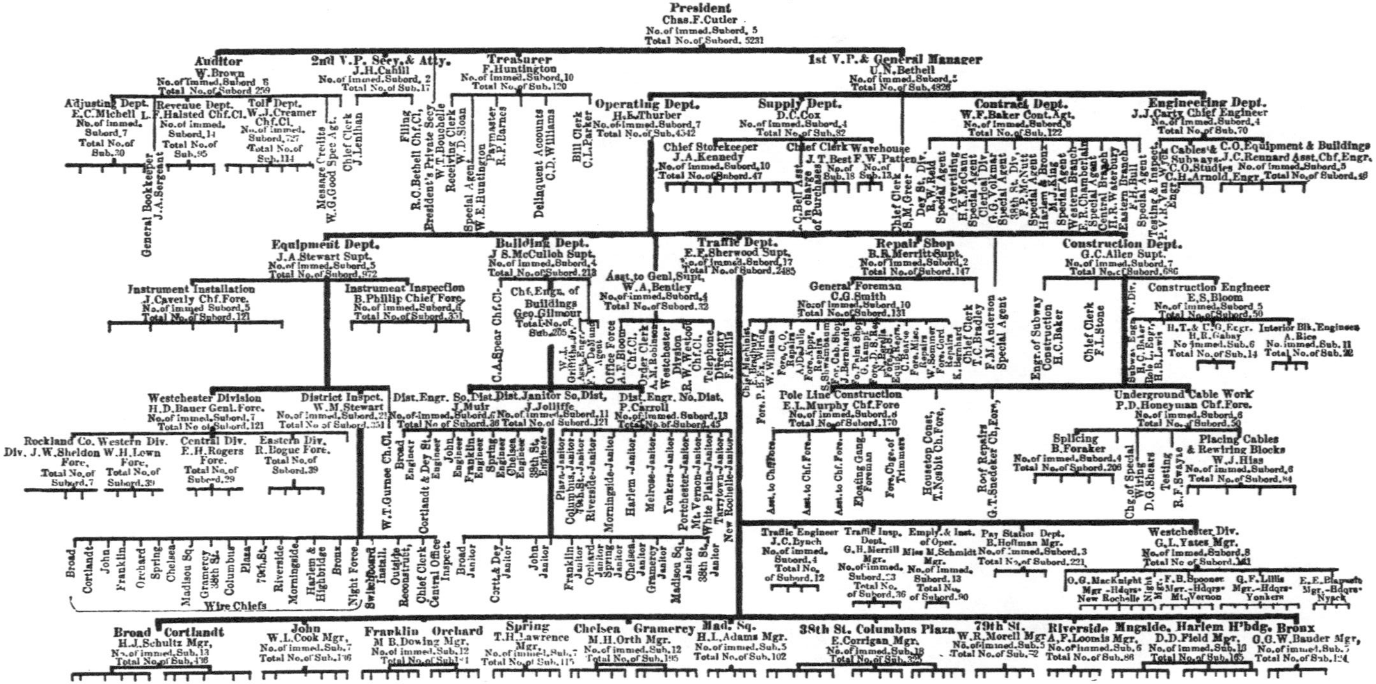

ice will be five, ten or fifteen years ahead. These studies are most carefully prepared and fit the future conditions with remarkable correctness, considering the uncertain nature of the factors involved. One very valuable aspect of this work lies in the training which it gives each department to make accurate predictions, and thus to allow adequate and timely provision to be made for future needs in any branch of the service. The chief engineer, who is at the head of this department, has a force of seventy men.

The Supply Department fills an important position in the organization. The purchasing agent is at the head of this department, and under his direction all supplies of whatever description are purchased and distributed. The electrical supplies necessary for the construction and maintenance of a large telephone plant are very expensive. Moreover, types of equipment are constantly changing, and it is necessary for the purchasing agent to keep in touch with the market at all times in order to be able to keep the telephone company supplied with the best material and equipment at a reasonably low price. The distribution of supplies is carefully looked after by the storekeepers, who are subordinate to the purchasing agent, and supplies are given out only on properly authorized requisitions. The force subordinate to the purchasing agent numbers eighty-two.

The Contract Department has to deal with the public in nearly all commercial relations. It makes all contracts for service and takes up all matters with subscribers, excepting those relating to the service, and the collection of revenue. This department keeps in touch with the business and social conditions of the city, and the needs of the individual and the public are studied, with a view to rendering in each case the service appropriate to the requirements of the user. The importance of this department can readily be seen, as it devolves upon it to supply the business which keeps the whole organization in action. The number of men subordinate to the contract agent is 125.

The Operating Department is responsible for the construction and maintenance, as well as the operation, of the entire plant. Under the general superintendent, who is the responsible head of this operating system, there are five active departments, all in close touch with one another, yet each dealing with an entirely distinct phase of the business. A brief description of these departments is as follows:

The Construction Department, under the superintendent of construction, as the responsible head, deals with what we may term the outdoor plant. It builds and maintains all underground and overhead construction, the sphere of its work being to deal with that part of the plant which connects the subscriber's interior equipment with the company's exchange. Although originally the wiring was entirely of the overhead type, a gradual reconstruction has taken place and the system is now entirely underground, excepting in a few of the smaller suburban towns, the proportion of the overhead wiring in Manhattan being only two per cent. In all underground work the engineering plans are carefully followed, and the "feeder cables" are laid with a sufficient supply of "spare circuits" to accommodate the steady increase in business and for use in emergencies. The initial expense of this underground cable work is large, but it is a very stable method of construction and contributes materially to the efficiency of the service. The total force of the superintendent of construction numbers 686.

The Equipment Department brings the subscriber's line from where construction leaves it to the actual point of operation. It equips the subscriber's end of the line and connects the same with the exchange on the other end. It installs and maintains all inside equipment and does all the inside wiring, and it makes a thorough inspection of the subscriber's equipment at least six times a year, while the central office equipment is constantly under the supervision of the wire-chief, who is an employee of this department. Perhaps the most interesting feature in connection with the work of the department is the remarkable way which it has of handling cases of "trouble." These "troubles" are divided into three classes, known as central office, line and station troubles. Statistics kept over a considerable period of time show that the average time of clearing a central office trouble is thirteen minutes; line trouble, two hours and fourteen minutes; station trouble, one hour and eight minutes. The plant is, however, so well built and maintained that these "troubles" are reduced to the minimum. The total number of employees subordinate to the superintendent of equipment is 972.

To provide for the housing of the inside plant as well as office and operating rooms, the New York Telephone Company has at present twenty-three buildings. The buildings are of the most modern fireproof construction, and from an engineering and architectural standpoint represent the best types attainable. They are planned not only with reference to the present switchboards which are housed in them, but are also so arranged that new switchboards can be added when desired and the old ones removed without interfering with the service. While

Public Pay Station operated by New York Telephone Company

the life of the switchboard is variable, the buildings themselves have an indefinite life and must be planned on such a flexible basis that they will permit of the removal of the old switchboards when necessary, and of the installation of new ones, and also, so far as possible, be adaptable to the various changes in the art which are constantly taking place. These buildings, after they have been constructed, are in charge of a superintendent of buildings, who is responsible for their proper care and maintenance. For this work he has a staff of 213 men.

The operation of the plant, the actual rendering of telephone service, is the work done by the Traffic Department. In point of numbers, this department is by far the largest, as it includes the operating forces of the seventeen Manhattan Exchanges and the twenty-seven exchanges scattered through the Borough of the Bronx and Westchester County, Rockland County and Connecticut.

This department also maintains in certain public places, such as railroad stations, ferry houses, hotels and large office buildings, public pay stations attended by regular employees of the telephone company. The total number of pay stations operated by the company is very large, Manhattan Island alone being supplied with over 5,000. So thickly are these pay stations dotted about over Manhattan Island that wherever one finds himself it is only necessary to take a few steps in order to reach a public telephone. This pay station service is of the utmost benefit to travelers and to the public at large, and is one of the features of the telephone system of New York which is universally appreciated.

An interesting auxiliary to the Traffic Department is the telephone operator's school. There, under experienced instructors, those applicants for positions as operators who have passed a searching preliminary examination are drilled in the principles and practice of telephone operating. The necessity for unfailing politeness and courtesy in all of their dealings with the telephone customers is a point which is insisted on from the beginning to the end of their work in the school. The training given in the school qualifies the operator for entrance to any of the company's offices, these offices all being operated on a standard method. By means

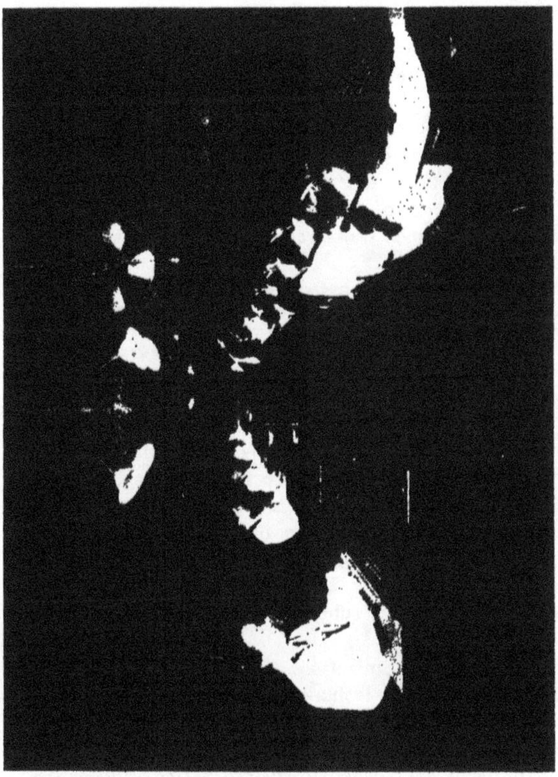

Telephone Operator's Training School

of this school the service of the subscribers is not impaired by the presence of unskilled operators at the working switchboards.

All the Central office employees, with the exception of the manager and his assistant, and the wire chief's force, are girls, as they are found to make the best operators. At the present time it takes about 2,100 of these operators to handle the New York business while at all times about ninety girls are in training at the school to fill the vacancies which are occurring daily throughout the ranks.

Careful handling of traffic is of vital importance to the success of the New York Telephone Company and every effort is made to render the most efficient service. A department of traffic inspection is maintained and every complaint regarding the service is taken up and carefully investigated, with a view to locating the trouble and preventing its recurrence.

The efficiency of the Traffic Department is shown in the time elements of traffic, taken from recent test sheets, as follows:

Average time of answering the calling subscriber, 3.8 seconds.

Average time of completing a local call, 30.7 seconds.

Average time of completing a suburban toll call, 58 seconds.

The personnel of the entire department is of a high order. This is indicated by the number of men with college training which are taken on each year to fill the various positions created by the growth of the company's business, and also by the fact that of the number of girls applying for operator's training in the school, only eight per cent. are accepted. The total number of employees subordinate to the superintendent of traffic is 2,485.

One of the organization developments incident to the growth of the telephone system is the Repair Department. This department which a few years ago consisted of but a dozen equipment repair men, working in a small repair shop, has grown into a large and active department occupying an entire new building on 18th street near Chelsea square, with the exception of the two upper floors, which are occupied by the new Chelsea exchange. This is a model workshop in every respect and is thoroughly equipped to do any kind of repair work which may arise in connection with telephone equipment. At the present time the superintendent of the repair shop has a total force of 147.

PRIVATE BRANCH EXCHANGES

An interesting development of the New York Telephone Company's service to meet the needs of a busy community is the private branch exchange. This is an application of the exchange principles to the subscriber's station. A switchboard is installed on the subscriber's premises, with as many lines to the company's nearest exchange as are needed, and local lines terminating in telephone stations throughout the building or off the premises wherever desired. This gives the large users of the telephone, such as hotels, apartment houses, department stores and large business firms, a most efficient and flexible service. The local operator receives and distributes the calls, both incoming and outgoing. There are at the present time in New York over 5,000 of these private branch exchanges, with a total of over 60,000 stations, and the number is constantly increasing. The value of this private branch exchange service to large users of the telephone can scarcely be estimated. To the large establishment it furnishes a complete interior service to all departments, yet at the same time each station is capable of immediate connection with any other station in the territory of the New York Telephone Company, and, in fact, any other station reached by the Long Distance wires. It means immediate accessibility at all times to any department and it solves the great problem of a large city, particularly in New York, of how to do things quickly.

The telephone traffic features presented in each section of New York City are individual. In the business district the tendency is to crowd the business day into as few hours as possible, but during those hours top-speed is maintained. This feature is brought out with clearness in the graphic chart showing the rise and fall of traffic in the Broad Exchange District. Practically all the business of this district emanates from large commercial exchanges and brokerage houses where the business hours extend only from 10 A. M. to 3 P. M.

In the theatre district a heavy traffic is continued

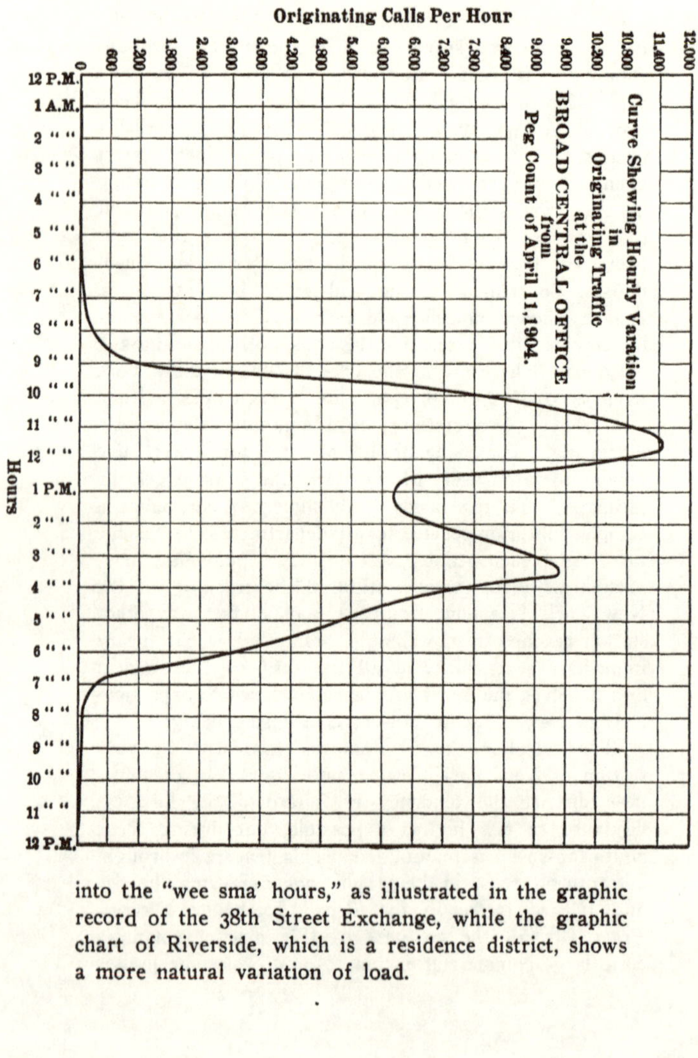

into the "wee sma' hours," as illustrated in the graphic record of the 38th Street Exchange, while the graphic chart of Riverside, which is a residence district, shows a more natural variation of load.

Manhattan Island has been divided into seventeen exchange districts, the facilities of each exchange being adapted to the special requirements of that region,

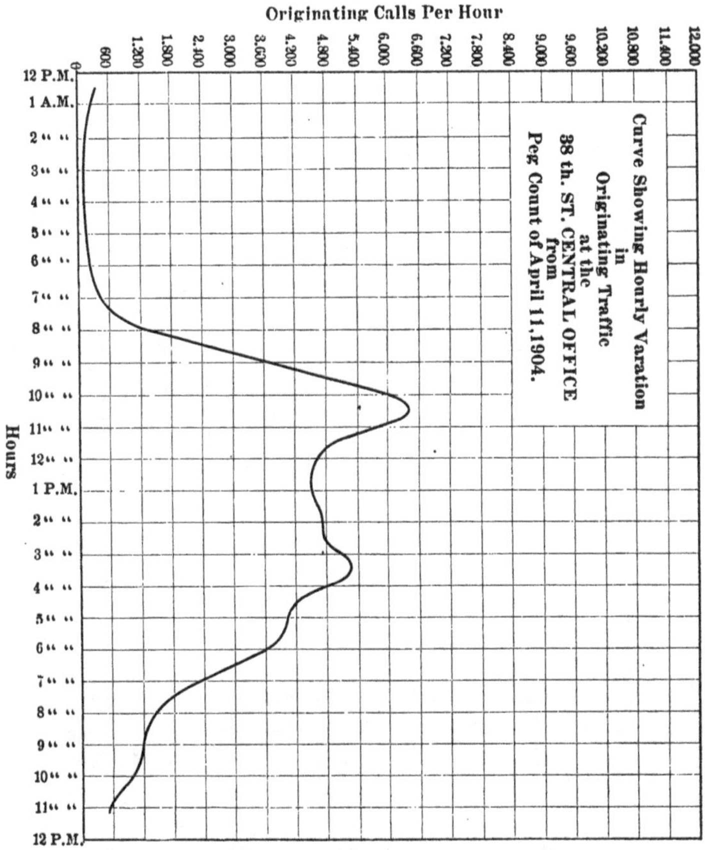

and also being ample to take care of the busiest hour of the busiest day of the busiest month.

To a high degree the service in New York must be efficient and permanent as well as rapid, and to effect this result the plant must not only be of the most modern construction, but it must be so protected and maintained that the interruption of the service in the slightest degree shall be an exceptional event. The avail-

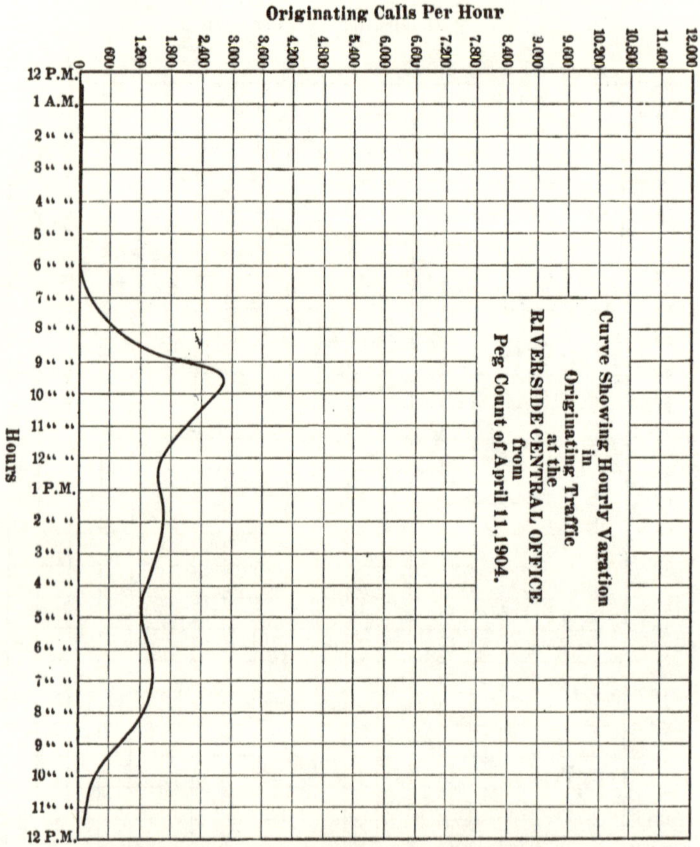

ability of the service must never be impaired for a moment of the twenty-four hours of any day of any year.

CENTRAL OFFICES

By far the most complex part of the telephone apparatus is the switching appliance in the central offices.

This switching apparatus is of a uniform type throughout all the central offices in the city and a description of one exchange will suffice for all.

The Cortlandt Exchange occupies an entire floor of the New York Telephone Company's building on

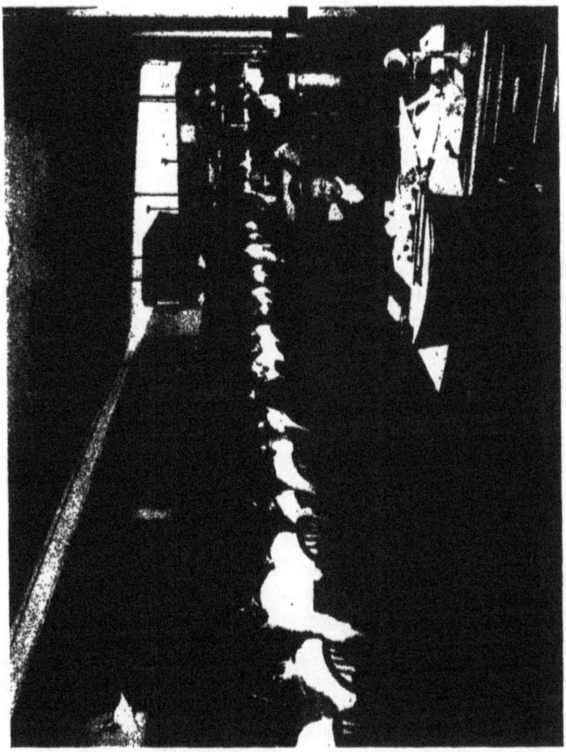

Cortlandt Street Exchange, north section of operating room

Cortlandt street and three floors of the adjacent wing, extending through to 15 Dey street. This wing contains the cable-terminals, distributing frames, storage batteries, power plant and other auxiliary apparatus.

The lines are brought in under ground in large paper-insulated cables. These cables are made up of twisted-pair copper wires, insulated with spiral wrappings of paper and the whole enclosed in a lead casing. The wires end in the main distributing frame and on the end of each pair is placed a carbon plate lightning arrester and a fuse. This takes care of any foreign current which may get on the line. The main distributing frame allows the changing of the entering wires of the subscriber's lines without changing the telephone numbers. The wires pass from the main frame to the intermediate frame, the object of which is to permit of any telephone call being answered at any position of the switchboard, thus equalizing the operator's load. These loads in the busy Cortlandt exchange range from fifty to ninety lines per operator. From this frame the line enters the switchboard and appears in an answering jack at some position and also appears in the multiple jacks in every section of the entire switchboard, known as the "A" or subscriber's switchboard.

When the subscriber takes his receiver off the hook, the exchange is signalled by the lighting of a small incandescent lamp placed immediately below the answering jack. The operator inserts one plug of her pair in the answering jack and this extinguishes the light. By means of a listening key, she connects the subscriber's circuit and finds out what the calling party wants; she then takes the other plug of the pair and connects the subscriber with the number asked for if it happens to be in her exchange. If not in her exchange, she connects to a trunk line leading to the exchange where the number is located.

Until the called party answers, the supervisory lamp associated with the calling plug remains lighted. When the subscriber answers, the lamp is put out. As long as these two lights are extinguished the subscribers have the receivers off the hooks and are using the line, when they hang up, the lamps light and this is the signal for the operator to disconnect. If the subscriber's line wanted should be in use, the operator will hear a sharp click when she plugs in on it; this shows her that the line is busy.

The calls from other exchanges coming in over the exchange trunk lines are received at the "B" or incom-

Back view showing section of telephone switchboard

ing trunk board. The operators at these boards simply put up the connections; they have no dealings with the subscriber.

The "A" switchboard of the Cortlandt exchange is 176 feet long and is divided into 30 sections. The "B" board is 82 feet long with 18 sections. The "A" board has a capacity for 9,000 lines with 840 outgoing trunks to the other exchanges. The "B" board has the same capacity for Cortlandt subscribers and 1,000 incoming trunks on separate boards makes a complete double track.

Wire Chief's desk with equipment for locating "trouble"

The switchboard contains over a half-million springjacks and switches and about 14,000 incandescent lamp signals. Each signal and springjack is connected to the wiring by several wires and all connections are soldered for greater security. The number of soldered connections in the exchange run well up into the millions.

Each operator is provided with her individual operating set, consisting of breast-plate transmitter, and head receiver, the operator connecting her set to the switchboard by a cord and plug arrangement. The number of operators in this exchange is 126.

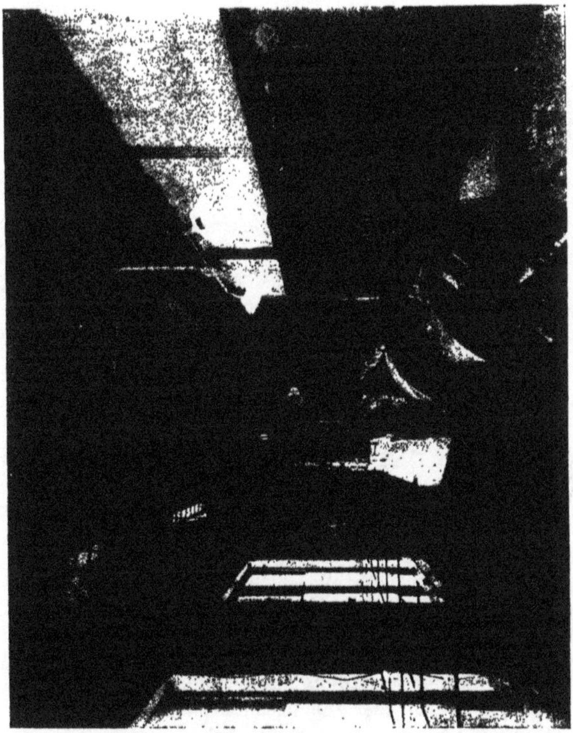

Power generators, storage batteries and cable runways, the latter leading from the intermediate distributing frame in the rear of the room to the switchboard on floor above

The entire ninth floor of the Dey street building is given over to the operators' quarters. Here is a spacious locker-room, with sanitary wire netting lockers, one for each operator; a large kitchen and dining-room where tea, coffee, etc., are provided by the company; a sitting room provided with newspapers and magazines; a sick-bay, lavatory, etc., all under the charge of a matron. These arrangements are provided in all the exchanges.

The maintenance of this part of the plant, the location of troubles, and the connecting and disconnecting of telephone stations, is carried on under the supervision of the wire-chief, who is an employee of the Equipment Department. He also has charge of the power plant, which occupies three floors of the Cortlandt street building. This exchange, as well as all others in New York, is operated on the common battery system, and in this exchange alone seven motor generators, aggregating nearly fifty-horse power in capacity, are used for furnishing current for charging the storage batteries and for the ringing and testing currents.

Every station in connection with any of the exchanges of The New York Telephone Company is equipped with a long distance telephone instrument with a "solid back" battery transmitter and a Bell receiver. These instruments are of the best type and are everywhere the recognized standard telephones. The wall set and the desk stand set are the two convenient styles in common use.

UNDERGROUND LINES

The wires leave the exchange under ground in lead covered cables. These are paper insulated and when the distance is short as many as 600 pair of wires are placed in a single cable. From these large cables in the subway the wires are distributed by smaller cables to the neighboring blocks. There are two conduit systems for carrying the underground wires in New York City, one containing the electric light and power cables, and the other the telephone, telegraph, and the other low tension wires. Each of these conduit systems is owned and operated by a separate company.

Although everything possible is done to guard against interruption of service, accidents will occur occasionally, the most serious being those due to fire. Recognizing this fact, the Construction Department has installed in the shop a fire-alarm and whenever a fire occurs an experienced lineman immediately looks up in the graphic records the exact location of the cables and wires on the premises threatened and immediately goes to the scene to look after the company's interests.

Since the subway work for the construction of the Underground Rapid Transit Company was started, the New York Telephone Company has been caused a

Cables at Broadway and Fulton Street before fire, manhole having been removed to make way for underground railway structure

great amount of serious trouble due to the disturbance of its cable plant. Although the subway fire at the corner of Broadway and Fulton street has been the only interference which has caused serious difficulties to the public at large, the continued disturbed condition of the streets for the past four years has given great annoyance, and the Construction Department has been obliged to keep constantly on the lookout to protect the plant as best it could in its exposed condition. The following incident will serve to show with what the Construction Department has had to contend:

To make way for the subway structure of the underground railway, in course of construction, the contractors removed the protecting manhole at Broadway and Fulton street provided for the cables, and massed these cables in a huge tangle. These cables were temporarily protected by being wrapped in burlap and surrounded by a wooden framework. By some unknown means a fire started in the excavation and was communicated to the woodwork and burlap around the cables. The heat, being intense, melted off the lead covering of twenty-two large telephone cables, including a Pupin long distance cable, connecting 5,250 pairs of wires, and fused them all together in a molten mass. Over 5,000 subscribers were instantly cut out of service.

After fire, cables brought to surface and spliced

In five minutes, picked men were dispatched to the scene to repair the damage. Within an hour and a half after the accident, the long distance wires in the Pupin cable were working, and in a little over twenty-four hours the entire system was working as if nothing had happened. The cost of this accident to the telephone company was over $10,000.

A peculiar electrical difficulty has always been presented in the operation of the underground plant. It has been found that to talk through a circuit of one mile of underground cable was as difficult as to talk through twenty-eight miles of overhead wire. Thus, a ten-mile stretch on Manhattan Island would present the same difficulty as 280 miles of overhead line. Telephone companies for years have experimented to overcome this difficulty, but the only invention which has helped to facilitate underground transmission is that of Doctor Pupin, of Columbia University. His invention consists in applying inductance to telephone lines. His method is to wind small coils of copper wire upon iron cores, and place one of these cores in each circuit at intervals of about a mile throughout its length. The matter of spacing these inductance cores was a subject of exhaustive mathematical study by Dr. Pupin, who determined that their successful use depended upon the number of them per telephone wave length.

The patents of Dr. Pupin's invention have been purchased by the Bell telephone interests, and the telephone engineers have been steadily at work perfecting the practical details of this appliance. This invention has been applied experimentally to the underground ends of the long distance lines branching out from New York, and satisfactory results have been obtained on these short lengths of cable. Further research into the extension of these Pupin cables is being made by the company's engineers with the promise of excellent results.

Still another phase of the cable work is that presented in laying wires across the Hudson and East rivers. The Borough of Manhattan being on an island, it is necessary to place wires across these rivers to establish connections with Long Island on one side and New Jersey on the other. For this submarine service special cables containing 200 pairs of wires are used. These cables are paper insulated in the usual way; then sheathed in a water-tight lead casing, and for further protection are also enclosed in a heavy wire armor. These cables are laid from a special

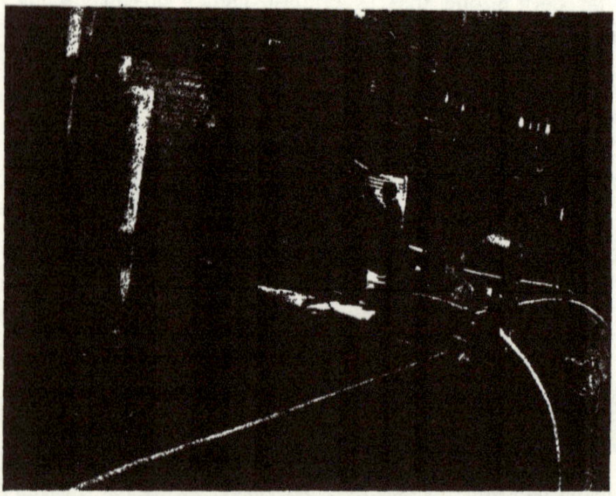

Cable tug laying a 30 pr. lead covered submarine cable under North River

cable tug, and are anchored in terminal houses on either side of the river.

The only disturbance of these cables which is in any way serious is that due to the fouling of the anchors of vessels. Although this feature cannot be obviated, it is taken into careful consideration when the cables are laid, and enough spare cables are placed to handle the service without interruption, notwithstanding the disturbed conditions due to this cause.

PLAN OF RATES

Up to 1894 there had always been a fixed rental for telephones, each subscriber leasing an instrument for one year and paying a fixed rate, regardless of use. It was found that this method of charging for telephone service worked an injustice to the customers whose use was relatively small, and since, at this time, the plant had been put in good condition, and it was possible to take on new business rapidly, the problem presented itself of readjusting the rate schedule so that the telephone service could be brought within the reach of all. In the solution of this problem the message was adopted as the unit of measure in the principal cities in the territory.

That the new plan was appreciated by the public is shown in the increase in the number of subscribers immediately following its adoption. At the close of 1894 there were 10,396 stations in old New York. At the end of 1897 there were 21,595 stations—an increase in three years of 11,199 stations. In 1903 the increase in Manhattan and Bronx alone was over 25,500 stations. In the entire territory of the New York Telephone Company and the New York and New Jersey Telephone Company there are now in service and under contract for immediate connection over 225,000 stations.

Since 1894 the rate plan has undergone a complete evolution, and every year that has passed has seen new changes in rate schedules. The object of this evolution has been to so adjust the rates and the forms of service that the telephone may be brought into the home in the city, or in the village, and may be utilized in all lines of business.

That rates are adjusted on a fair and reasonable basis throughout the territory is shown by the facts that there is a continual gain of stations and that there are more telephones in this territory than in any other similar territory of the world four times its size.

In the development of this system and in the construction and operation of the plant extraordinary difficulties have had to be overcome. Thirty years ago the telephone was a new invention and the public was skeptical as to its practical value. It was not only necessary to perfect an organization and build a new plant, but also to convince the public of the utility of telephone service.

In 1888, when the system was beginning to be fairly well established in New York, the construction was of the overhead type and the streets were full of overhead wires. At this time the subway laws were passed and a most serious problem was presented; the overhead lines must be abandoned and an underground system constructed to take their place.

There was no chance of benefiting by the experience of others, as there were no other cities either here or abroad where similar work had been done to any great extent and after expensive experimental work, new central offices were constructed and a new method of working the entire system was determined upon.

Since 1886, the entire plant has been reconstructed three times, in order that by bringing the equipment up to the highest possible standard, the service might be rendered more efficient,. and it can readily be seen that the expense of bringing this plant up to its present standard has been enormous.

THE METROPOLITAN STREET RAILWAY SYSTEM

The Metropolitan Street Railway System

THE Metropolitan Street Railway system embraces all the street surface railway lines of the Boroughs of Manhattan and the Bronx. Its lines in the Bronx also extend widely into the adjoining county of Westchester, its northerly termini in the villages of Tarrytown, White Plains and Mamaroneck being distant about thirty-three miles from the southern extremity of the system at the Battery. Within these limits, with about 500 miles of track, it serves a resident population of nearly two and one-half millions, of which all but about four hundred thousand are south of the Harlem River. It traverses the most densely populated districts in the world and its traffic and earnings exceed those of any other street railway.

The system is separated into two general divisions by the Harlem River, the lines north of the river being operated independently of those in the Borough of Manhattan. These two divisions differ radically in characteristics. The system in Manhattan is essentially and completely urban, with dense traffic on nearly every line. Its electric construction is all of the conduit type, substantial and costly in the extreme. In the territory of the Union Railway Company (the company which operates the lines north of the Harlem River) totally different conditions prevail. Its lines extend from termini in the Borough of Manhattan across the Harlem River bridges into sparsely populated districts of the Bronx and Westchester county, and its traffic is largely suburban. It is operated exclusively by the overhead trolley and many of its outlying and unimportant lines are still single track.

HISTORY

The Metropolitan system is the result of successive leases, mergers and amalgamations of over forty independent companies, organized under a great variety of charters, scarcely any two of which are similar in their provisions. A brief history of the inception and growth of the street railway enterprises now vested in the New York City Railway Company is worthy of passing notice. It commences in 1832 with the building by the New York & Harlem Railroad Company of the first street railway line in the world. This line extended from City Hall Square to the village of Harlem over a route which is now embraced in the Fourth Avenue line of the Metropolitan system and the present tunnel and viaduct of the New York Central & Hudson River Railroad Company. This line, which at first was operated by horses, was later extended across the Harlem River and the horse cars replaced by steam locomotives and trains. The growth of the city in later years caused successive withdrawals of steam service, first, in 1854, below what is now Madison Square Garden, and, again in 1870, below the Grand Central Station.

In 1835 the first stage line for local travel was opened over the route of the present Third Avenue line between the City Hall and Harlem, but it was not until 1852 that there was any addition to the street railway mileage. With the opening in that year of the Sixth Avenue Railroad, with four miles of line, development became active, and in 1854 the Third Avenue Railroad Company, organized a year previous to take over the stage line operating on that thoroughfare, added four miles more, closely followed by the Second Avenue Railroad Company and the Eighth Avenue Railroad Company, each with about the same mileage. By 1860 there were sixty miles of horse car lines in operation on the Island of Manhattan, and by 1870 this had increased to 142 miles.

Between 1870 and 1880 the elevated railroads were constructed and opened, making serious inroads on the traffic of the surface lines—so much so that many of the owners of the latter feared they would never again pay operating expenses. In 1885 a franchise was granted for a railroad on Broadway from the Battery to Central Park. This famous street had up to that time been served only by stage lines and the opposition to the construction of a railroad was very great, predictions being freely made that the addition of cars to the vehicle traffic would congest the street to such an extent that it would be ruined. Although at that time cable railways were not uncommon and the Third Avenue Railroad Company was installing such a system on its 125th Street and Amsterdam Avenue lines, the Broadway road was built for horse operation, and as such proved the fastest line in the city and upset all the predictions of its opponents by facilitating travel on that street in a manner unknown theretofore.

About this time a group of Philadelphia capitalists conceived the idea of acquiring and consolidating the street railways of New York. They acquired first the Chambers Street & Grand Street Ferry Railroad Company and later the Houston, West Street and Pavonia Ferry Railroad Company, the consolidation of these two lines being the first step in the merger. Strong financial interests from New York entered the syndicate, comprehensive plans were made and in rapid succession the Broadway and Seventh Avenue, Sixth Avenue, Ninth Avenue and 23rd Street roads were leased and added to the system. From that time to 1900 all the competing and independent properties on the island of Manhattan were one by one acquired by purchase or lease. In that year the last step in the consolidation was taken in the bringing into the sytem of the Metropolitan Company's only surviving competitor, the Third Avenue Railroad Company, with seventy-one miles of track in the Borough of Manhattan and one hundred and thirty-five miles north of the Harlem River, together with unconstructed franchises of great extent and value.

The final organization of the several corporations through which the properties are now controlled is a little complex. It suffices to say that the Metropolitan Securities Company is the controlling organization and that the New York City Railway Company is the operating company south, and the Union Railway Company north, of the Harlem River. The organization of the New York City Railway Company for operating purposes is shown in the accompanying chart. The Union Railway Company has its own executive and operating organization and the officers of the New York City Railway Company take no part in its management.

In the matter that follows, the Union Railway Company should not be understood to be included unless specifically so stated, as the space allotted to this article will not permit a description of its system which presents no features differing greatly from those of many other overhead trolley roads in the United States.

INTRODUCTION OF MECHANICAL TRACTION

The earliest plans of the syndicate for the consolidation of the New York roads contemplated the application of an improved motive power to the entire system. The overhead trolley was coming rapidly into use as the most approved form of surface traction, and had been adopted by the West End Street Railway Company of Boston for its entire system of 250 miles of track—at that time the largest city railway system in the world. In September, 1892, a portion of the Union Railway (then an independent road) was equipped with the overhead trolley—the first electric railway mileage in the City of New York. But New York was at that time extensively engaged in burying overhead wires of all kinds and, for the Island of Manhattan, at least, the opposition to the overhead trolley, both for aesthetic and practical reasons, was so great as to be prohibitive. Philadelphia, Washington, Chicago, San Francisco and other cities had the cable in successful operation—in fact, it was already in use on the Third Avenue system in New York; and for the Broadway line, which had developed what was then an enormous traffic, continuous throughout the day, with a short, practically straight

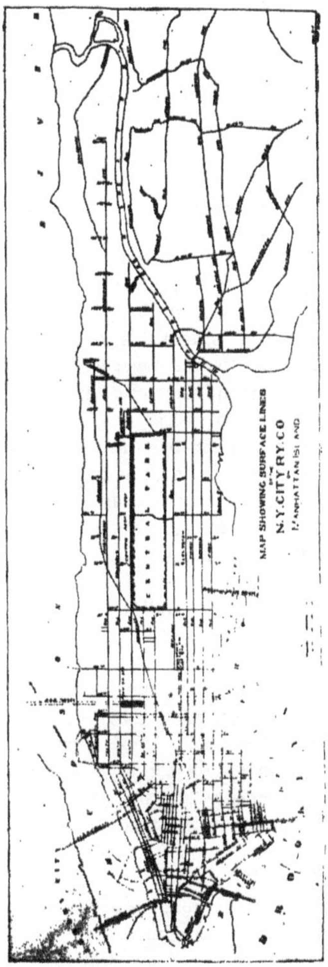

Map showing Surface Lines of New York City Railway Company

run, it seemed the best power available and was duly adopted. The opening of the line from the Battery to Central Park in July, 1893, was attended with such success that plans were immediately made for extensions to Harlem on either side of Central Park, through Columbus and Lexington Avenues. The cable system was, however, so enormously expensive both to construct and operate, and was capable of so little flexibility, that it was recognized from the first that its application must be confined to lines of the first importance, and that a more flexible and economical system must be found. The Company's engineers were sent to Buda-Pesth, Hungary, and other European cities to study the latest development, and a reward of $50,000 was offered for the submission of a practical system of motive power equal or superior to the overhead trolley. Nothing practical resulted from this offer, and early in 1894 the Company proceeded to adapt to its uses a conduit system similar to that which had been in use in Buda-Pesth for several years. The Lenox Avenue line in Harlem, three miles in length, designed as a continuation of the new cable lines on either side of Central Park, was selected for the experiment, which was conducted on the most conservative lines. In order to reduce to a minimum the chances of failure, the construction was planned as for a cable road so that, should the electric system prove unsatisfactory, the cable could be substituted. In July, 1895, the road was opened for service and proved so efficient, reliable and economical in operation that the extension of the electric conduit system was decided on, not only for all new lines, but for the ultimate displacement of the cable, over which it showed an economy in operation of at least five cents per car mile.

The problem of providing a system as low in cost of construction as the overhead trolley, and which could be applied with profit to all the many miles of horse car lines, had not yet been solved, and experimental work costing large amounts continued to be made for several years. A surface contact system was tried and found wanting. Compressed air seemed to offer greater possibilities than any other self-contained motor, and after several preliminary trials the 28th and 29th Streets line was selected for a test of that power, and, in August, 1899, was completely equipped and opened for operation. A large compressing and storage plant with a 1,000-H.P. direct-driven, four-stage compressor, capable of compressing to 2,400 pounds to the square inch, and with a capacity of eighty cars, was built. The results extending over nearly two years and under close observation were very unsatisfactory, and in April, 1901, following a succession of failures, accidents, irregularities in service and general dissatisfaction on the part of the public and the city authorities, the air car service was withdrawn and horses restored.

An experimental trial quite as thorough and extensive, and more satisfactory in result was made of the chloride storage battery. In August, 1900, a large battery station with a capacity of fifty cars was established at the foot of West 42nd Street for the operation of the 34th Street line, and from that time until September, 1903, the line was operated by the storage battery cars without serious trouble or interruption. Failures were frequent, however, discomfort was occasioned by the fuming of the batteries, and the cost of operation was very high, so that, during the summer of 1903, the conduit construction was installed, the traffic of the line having increased to such an extent as to warrant the improvement.

PERMANENT WAY AND SUBSTRUCTURE

As stated above, the installation of the conduit system on Manhattan Island was begun in 1895. Additions have been made to the mileage so constructed each year, until at the present time there are two hundred miles of track so equipped.

The system is known as a double conductor system with metallic return circuit, and no use is made of the tram rails for returning current to the power stations, and was adopted in practically its present form, no changes worthy of note having been found desirable beyond the strengthening of all the component parts of the plow.

This form of construction has both advantages and disadvantages, as compared with the overhead trolley system. All the construction being underground, the feeder wires and electric conductors are protected from damage by wind and weather, and the expense of maintenance of the electric conductors is somewhat reduced as compared with the overhead trolley system. On the other hand, the cleaning of dirt from the conduits and the manholes is an added expense, while the presence of the centre slot complicates the construction of special track work and increases the expense of its installation and maintenance.

The absence of poles and overhead wires, always more or less unsightly, is a distinct advantage from an æsthetic point of view, while possible danger to the public from falling wires is avoided.

Proper drainage of the conduit and manholes is obtained by connecting each manhole with the street sewer by a six-inch pipe. These manholes being only about one hundred feet apart, this provision easily takes care of the surface water which may find its way into the conduit through the slot. Reference to the cuts gives a fair idea of the arrangement of the conduits and their position with reference to the surface of the street.

The conduits are cleaned by drawing through them a scraper, which conforms to the cross-section of the conduit, and which is carried by a car running on the track above. This scrapes such dirt as may have fallen through the slot along the conduit, depositing it in the handholes and manholes, from which it is afterwards removed by the cleaning gang and taken to the city dumping places.

There are at present about 160 men engaged solely in this work of keeping the underground construction free from dirt. Such cleaning of the conduit costs about $500 a year per mile of conduit.

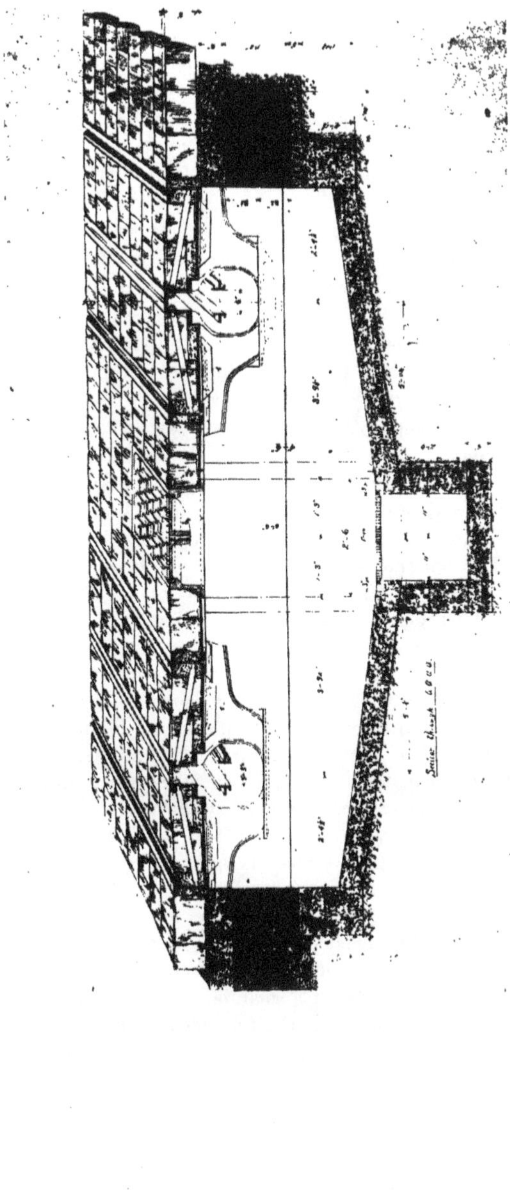

Section through Man Holes

At certain places on several of the lines the level of the conduit is so low that at extreme high tide the conductor bars are covered with water and the operation of the cars over that part of the line is necessarily suspended until the water level recedes. This sometimes happens during very heavy rains, when the city sewers in certain locations are not sufficient to take the surface water away from the streets as fast as it falls. This occurs perhaps once or twice a year.

Considerable trouble has been experienced during severe winters, like the last, because ice and snow collects in the conduit and manholes and in handholes around the insulators. This trouble is especially severe when the snow and ice is frozen in the handholes, and manifests itself in broken rail brackets, grounded insulators and a variety of other ways.

The Company's facilities and the construction of the conduit are such that any ordinary snow fall up to ten or twelve inches in depth can be easily handled, and experience shows that the conduit system is fully as reliable in its operation as the overhead system.

The pavement used in the Borough of Manhattan in connection with the reconstruction of the electric lines is largely asphalt. Some granite block pavement is in use, and a little wooden block pavement. From the standpoint of the operating Company, the granite block or the wooden block pavement is to be preferred to asphalt, as either of these pavements can be more easily maintained in good condition along and between the rails than can the asphalt, which quickly cuts out leaving the street surface in a very undesirable condition.

The construction of this style of conduit trolley road in New York City streets is quite difficult, for many reasons.

In the first place, the volume of street traffic on nearly all the lines which have been constructed is large, especially in the lower parts of the city, and, in addition to provision for maintaining the operation of the horse cars until the electric cars are ready for busi-

Underground Obstructions

ness, arrangements must be made for the continuance at all times of the ordinary street traffic.

Furthermore, in any street in New York, the surface having been removed, one encounters divers obstacles in the shape of gas pipes, water pipes and electric conductors, laced and interlaced, over the entire street, and so near the surface that a rearrangement of many, if not all, of these structures is necessary in order that a free passage for the conduit construction may be provided.

The cost of construction of conduit roads, as compared with the track and special work required with overhead trolley lines, is very much in favor of the latter. In fact, the average cost of conduit construction per mile of single track in New York City is somewhat above $100,000 per mile; while the cost of special work at intersections, etc., is probably three times what it would be for use with the overhead trolley. The reason for this is found in the slotted construction necessary to provide access to the conduit. The life and the expense of maintenance of the *straight* track conduit construction does not appreciably differ from that of the overhead trolley construction, but the life of the special work does not average on the system of the New York City Railway Company more than five or six years, and the depreciation charged to maintenance is correspondingly high.

In general, the methods pursued in the conversion of the horsecar lines to electric lines have been, first, to provide for the continuance of the operation of the horsecars during the electric construction, either by laying a side track along the whole length of the line to be constructed, and operating the horsecars on this side track and on one of the main tracks while the other main track is being built, or else to remove the horse cars entirely to a line in some parallel street while the work of reconstruction is going on.

The necessary excavation is then made as deep as is required for the yokes and conduit of the new construction. This exposes such water pipes, gas pipes, electric

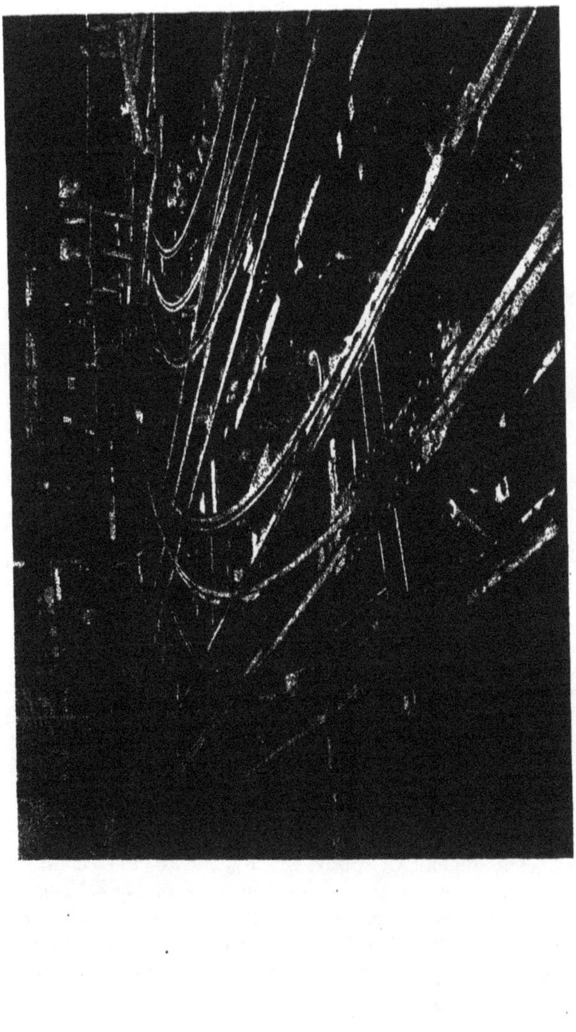

Special Track Work

conductors, etc., as may interefere with construction. Such pipes are then moved to positions where they will not interefere with the construction.

Then the iron yokes, steel rails, etc., used in the construction are put in place and bolted up and the track lined and levelled and temporarily supported, while Portland cement concrete is rammed beneath and around the yokes supporting them and forming the conduit, which later will contain the electric conductors. At the same time, the terra cotta ducts for containing the electric cables are put in place, and lastly the pavement—asphalt or granite blocks—is put in place and the construction is ready for the electric conductors, the installation of which is the last phase of the construction.

The conversion of the cable lines to electric lines was done in substantially this manner, except that it was necessary to keep the cable cars in continuous operation while the change was being made. This was done in such a manner that when the time came for the final change from cable power to electric power, the operation of the Broadway line was discontinued at nine o'clock on Saturday night, and on Sunday afternoon at five o'clock the operation of electric cars was begun. The same method was pursued on the other cable lines as upon the Broadway line.

The miles of electric conduit road constructed by the Company in 1897 and yearly thereafter to date is given below:

1897..........	28	miles single track
1898..........	62	do
1899..........	44	do
1900..........	4	do
1901..........	33	do
1902..........	14	do
1903..........	15	do

The amount of special track work at intersection of lines and crossovers, etc., in the City of New York is excessive, as compared with the number of miles of straight track. There are now in operation about twelve hundred switches on two hundred miles of straight track. These switches are arranged to be thrown either by switchmen employed for that purpose or by motormen or conductors on the cars.

Ninety-Sixth Street Power Station

In this connection, experiments have been carried on within the last year with two or three different styles of automatic switches which are thrown electrically and governed by the motorman of the car approaching the switch. These experiments have been quite successful and will doubtless result in the installation of quite a number of these switches.

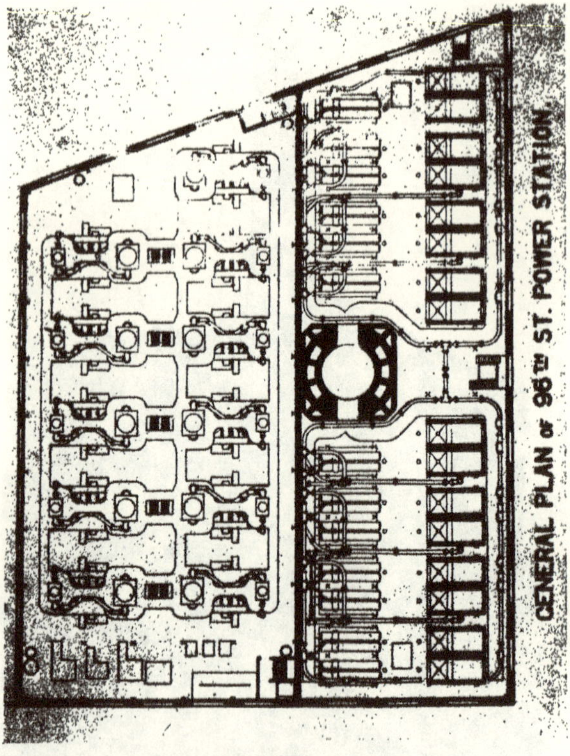

POWER

A glance at the map showing the electrically equipped surface lines on the Island of Manhattan shows how difficult and costly would be the task of supplying the electric current to operate all these lines from direct current power stations. A careful study of the conditions led to the adoption of the plan of the central generating station transmitting alternating current at high voltage to substations located at different points throughout the city, where the high tension alternating current is transformed to low tension direct current and distributed to the cars throughout the limited area covered by each substation.

This general plan has been carried out with slight modification, and the New York City Railway Company has on Manhattan Island to-day two generating stations, each generating current at 6,600 volts and transmitting it to various substations not only on Manhattan Island, but also in the Borough of the Bronx and in the cities of Yonkers and Mt. Vernon.

One of these generating stations is located at 96th Street and the East River; the other at 218th Street and the Harlem River.

The 96th Street Power Station of the Company was the first of the several large polyphase power generating stations to be erected and operated in the City of New York. Construction was begun on foundations in 1897, and in August, 1899, the first unit was put in operation.

The station contains eleven 3,500 k.w generating units. The current generated is three-phase, twenty-five cycle, 6,600 volts, and is distributed at that pressure to seven substations on the Island of Manhattan.

The electric generators are driven by vertical cross-compound condensing engines directly connected to the generators. Surface condensers are used for each engine, and all station auxiliaries are steam driven.

Steam is furnished to the engines by eighty boilers of the inclined water tube type. These boilers are arranged on the first, second and third floors of the boiler house. They are set in batteries of two, each battery being rated at 500 H.P. nominal.

The coal is taken from the boats by steam shovel and automatic conveyors to the coal bunkers, which are located just under the roof of the boiler house, and which have a capacity of 9,000 tons. From the bunkers the coal descends by gravity to the automatic stokers, with which all the boilers are equipped, and by them is fed directly to the fires.

The output of this station has been in times of heavi-

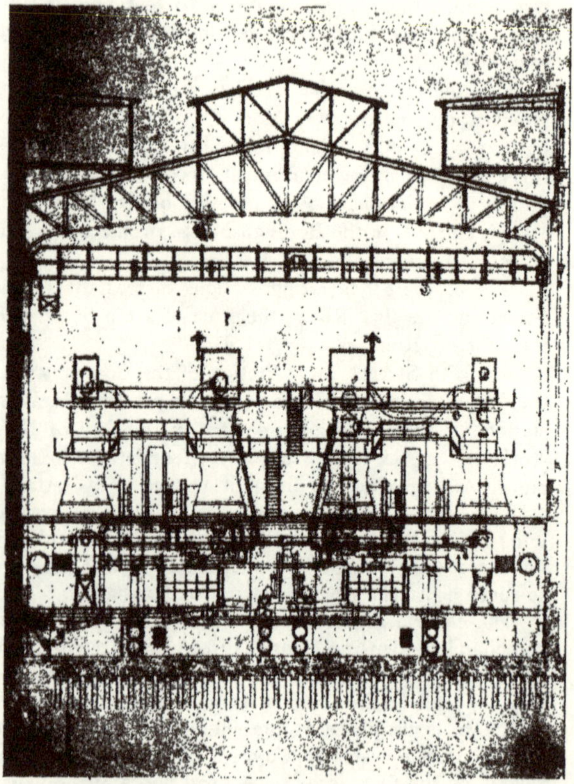

Section Engine Room 96th Street Power Station

est load—which occurs about Christmas—as high as 575,000 kilowatt hours per day. The output during the maximum hour of the day has frequently gone as high as 36,000 kilowatts, or about the nominal load for the eleven generating units.

A typical load curve of this power station is shown in the cut. The load factor of the station, of course, varies somewhat with the season of the year, varying from sixty per cent. to sixty-six per cent.

The coal consumption at this power station is about

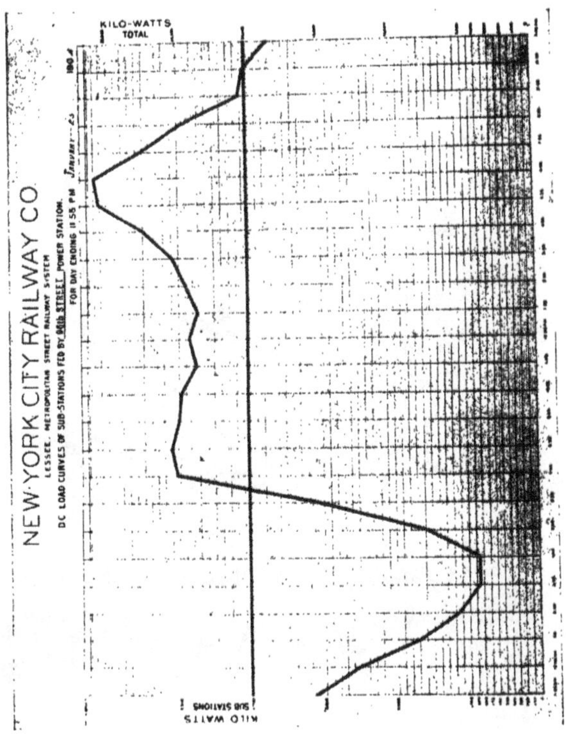

two and eight-tenths pounds of coal per kilowatt. The coal consumption for the maximum output above mentioned would, therefore, be about 718 tons per day. The total output of the station for the year ending December 31 last amounted to over 149,000,000 k.w.h.

In order to ensure continuity of operation, and to

provide as far as possible against interruption due to disabling of any piece of apparatus, the station is arranged so that it can be operated as three independent

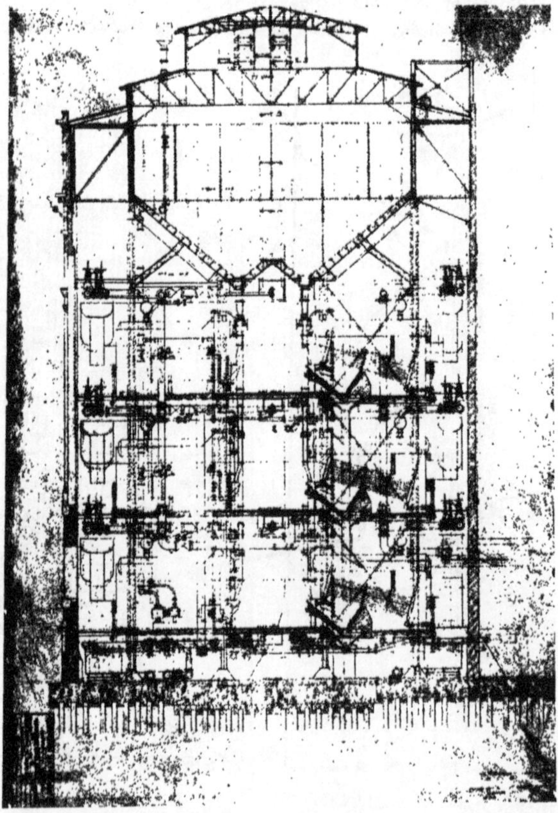

Section Boiler Room 96th Street Power Station

stations if desired. It has not been found necessary to make this subdivision, however, and the station has thus far been operated as one unit. In the last four years of operation there have been but two interruptions to

Kingsbridge Power Station

the steady delivery of electric current from this power station, and in both of these instances the trouble was due to causes outside of the power station itself.

A reference to the several cuts of this power station shows how very compact is the arrangement of the machinery. As a matter of fact, the ground space required per k.w. of capacity in the 96th Street Power House is 1.24 sq. ft. for boiler and engine rooms together. Of this .68 sq. ft. is in the engine rooms, and .56 sq. ft. in the boiler room.

The Kingsbridge Power Station, situated at 218th Street and the Harlem River, supplies current to the two northern substations on Manhattan Island; to the Borough of the Bronx, as before stated, and to the substations at Mt. Vernon and Yonkers.

The Kingsbridge Power Station has only recently been finished, the first unit having been put in operation less than one year ago.

In arrangement this station follows the same general plan as the 96th Street Power Station. The electric units are of the same size, 3,500 k.w. each, and generating three-phase current at twenty-five cycles and 6,600 volts.

The generating units are eight in number and are driven by vertical cross-compound condensing engines directly connected to the generators.

The steam is supplied by water tube boilers of the inclined tube type. These boilers, thirty-two in number, are arranged on the first and second floors of the boiler house and are set in batteries of 1,000 H.P. each.

The coal bunkers are in the top of the boiler room, the coal being delivered to them by conveyors and fed from the bunkers by gravity to the automatic stokers with which the boilers are equipped.

The condensing plant of this station is of the type known as the central jet condensing, and is arranged in duplicate, as are all of the station auxiliaries. All auxiliaries are steam driven.

This station is not at present, and never has been, fully loaded. The space required per k.w. of capacity in this station is 1.16 sq. ft. for boiler and engine rooms

Engine Room Kingsbridge Power Station

together. Of this .58 sq. ft. is in the engine room and .58 sq. ft. in the boiler room.

In the Kingsbridge Power Station, as well as the 96th Street Power Station, the arrangement of the apparatus is such that they are practically several power stations under one roof. This subdivision holds from boiler room to switchboard. While the separation of the various groups of apparatus in the power house may be complete, if necessary, in ordinary operation all are connected together.

In these two power houses the Railway Company has ample power provision for any possible requirements for several years to come.

The mechanical force in the power stations is directly in charge of the Mechanical Engineer of the Company, and the electrical force is in charge of the Electrical Engineer; both of whom report to the Chief Engineer of the Company.

Each power station has a First, Second and Third Assistant Engineer, who stand watches of eight hours each, and are in direct charge of the operation of the power house during the eight hours. Under them are watch engineers, oilers, pump men, stokers, coal passers, water tenders, and so on; all working in eight-hour shifts.

The high tension feeders laid from the power stations to the substations are arranged in groups so that no substation is dependent upon current from any one feeder, as there are always several feeders connecting each substation with the power house. These feeders take different routes through the streets and, entering the power house from different directions, terminate in different sections of the feeder board.

This subdivision of the power station, and the feeders therefrom, provides all the practical advantages of several separate power stations. Any mechanical or electrical trouble with any of the power-house apparatus can hardly extend further than the group in which it originates, except temporarily.

The possibility of fire is practically eliminated by the

Interior West Farms Station

fireproof construction employed throughout the power stations.

The coal used at these power stations is partly anthracite, buckwheat size, approximately, 12,000 B. T. U.'s per lb., and partly semi-bituminous coal having a heating value of 14,500 B. T. U.'s per lb.

The average amount of water evaporated per pound of coal throughout the twenty-four hours, as measured by water meters, is about eight and one-half pounds. Water required per kw. hour at the switchboard is about twenty-two pounds, including all the auxiliaries about the power station.

The cost of coal is sixty-seven per cent., the cost of repairs to power plant is seven and six-tenths per cent., and the cost of labor is nineteen per cent. of the total operating and maintenance cost; while the yearly cost of repairs is about one and sixty-one-one hundredths per cent. of the first cost of the plant.

The location of both these power stations on the river front is such that coal is delivered to them directly by boats ensuring a proper supply of fuel at the lowest possible cost for handling, and ample water is available for condensing purposes.

The substations throughout the city have been located on property owned by the Company. In most instances the substation building is part of a car house. It so happens that the property owned by the Company is so situated that the substations are not far removed from their theoretically proper positions.

The electrical conductors in the conduit are divided into sections isolated from each other so that difficulties in the way of electrical grounds, etc., are confined to the section in which they occur. These sections average, approximately, one-half mile lengths, and each has its own set of feeders leading directly to it from the substation. This localizes trouble on the lines and facilitates quick repairs.

Switches are provided—located in suitable boxes in manholes—at the points of division of various sections, so that in case feeder trouble should temporarily disable traffic on one section, power can be quickly furnished to the disabled section by connecting it through the switch to the adjacent section. It has been found that this provision against power failure is so effective that delays of more than four or five minutes to cars through feeder troubles are quite rare.

The electrical troubles met with in operating this system can be roughly grouped under three heads: Troubles with feeders between substations and conductor bars; troubles with the conductor bars themselves, and troubles with electrical equipment of cars.

Nine-tenths of the feeder troubles occur at the manholes and are due to some mechanical injury to the feeders by men while working in the manholes, or by men working in other excavations alongside the company's duct line. It sometimes happens that in manholes containing a great many feeders a feeder low down in the rack may short-circuit or ground to the lead cover, from some cause or other, and burn not only itself, but other feeders above it until every feeder in the manhole is more or less damaged; the attendant at the substation meanwhile knowing nothing of all this until advised from the outside by the Inspector.

Troubles with conductor bars come mainly from three causes:

1st. Short-circuiting of plows or current collectors carried by the car. When this occurs the conductor rails are burned and buckle from the heat developed by the burning plow. Such rails must be removed and new ones put in their places.

2nd. Troubles due to accumulations of snow or ice in the conduit; and

3rd. Troubles caused by boys putting metallic articles, wire, chain, etc., through the slot and thereby short-circuiting the conductor bars.

The methods of prevention of troubles from the last two-named sources are obvious. The trouble from the first cause can never be entirely prevented, but only mitigated.

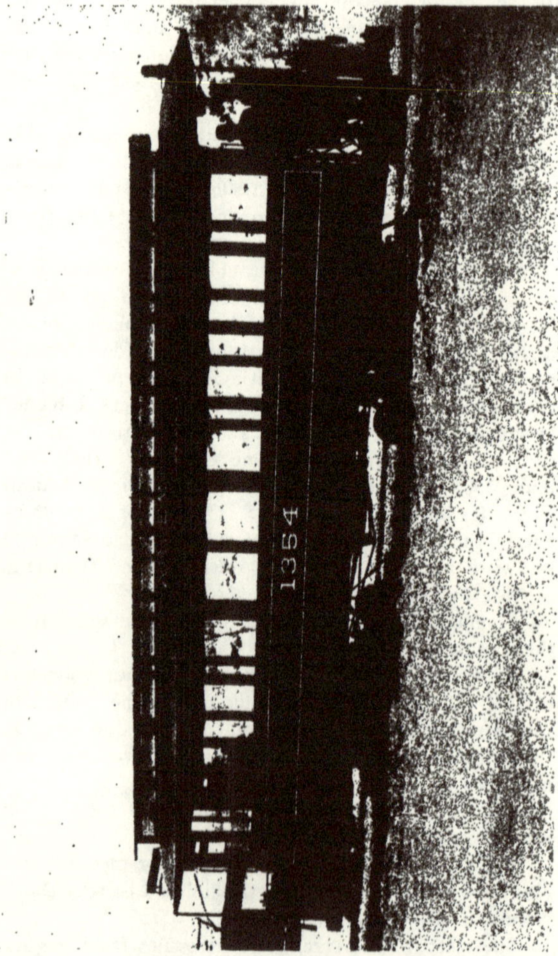

Standard Closed Car

For the proper care of the feeders and electric conductors in the conduit the city is divided into eight sections, and all the cable and electric conductors in each section are in the charge of a section foreman who has a force of men under his control and who is responsible directly to the Superintendent of Lines and Feeders for the condition of the electric cables and conductor bars in his section, and for the proper handling of his men. The Superintendent of Lines and Feeders reports directly to the Electrical Engineer.

ROLLING STOCK

The car equipment in use on the Metropolitan system cannot be described without, as in the description of other features of the system, referring to the history of the past eleven years with its evolutions from horse to cable and cable to electric power.

The first cable cars, built in 1893 for the Broadway Road, were twenty-two feet in body, thirty feet six inches over all, on a four-wheel rigid truck. These cars were all converted to electrics in 1901, when the power was changed on Broadway and Lexington Avenue and Columbus Avenue, and are in use in various parts of the system. The same thing is true of the cable cars belonging to the Third Avenue Railroad Company, which came into the system with the lease of that road in 1900. The first electric cars built as such for the experimental road on Lenox Avenue in 1895 were of the same size, equipped with the same truck and driven by two G. E. 800 motors. From these three sources came most of the "short" cars found on the system. With the definite adoption of the electric conduit system in 1898, the double-truck car, twenty-eight feet in body and thirty-six feet over all, mounted on "maximum traction" trucks, became the standard, being, at that time, a distinct departure from the prevailing practice. The object sought in the design was to obtain the largest unit that could be handled by two men and propelled by two motors, with the largest

Standard Open Car

possible seating capacity consistent with rapid ingress and egress. Limits on the size of this car were also set by the heavy trucking along New York streets, by the narrowness of these streets, by the sharp curves, which are in many places unavoidable, and by the Elevated Railroad pillars which are oftentimes badly placed with reference to these curves. Even where there are few difficulties of this kind in the large north and south avenues, there is sometimes a limit to the width and length on account of the narrowness of a few cross streets through which the line in some part of its route is obliged to pass.

In the early days of mechanical traction in New York the open car was not used through fear of accidents to passengers from passing vehicles, but the demand from the public became so pressing that it was finally acceded to and the open car proved exceedingly popular. A large number are in service on all the longitudinal lines. By the use of the side bar and strict prohibition of any standing on the foot-board the excessive danger predicted has not been realized, though the accident record is in favor of the closed car. So popular are the open cars in bright, warm weather, that closed cars operated on the same line with them pass practically empty, while the open cars are crowded. To meet this demand and, at the same time comply with the requirements of the Health Department, which call for one closed car in every five cars operated, a combination open and closed car, seating thirty-five passengers in the open part and sixteen in the closed compartment, was adopted and is in use on most of the longitudinal lines. This particular type has, however, proved objectionable in some respects and is not being included in current orders for new equipment.

The plow or underground trolley is confined to New York and Washington among American cities. The cast iron shoes used in the plow last from ten days to two weeks. At the line of the slot-rails there are inserted in the main plate two hardened steel clips called wear-

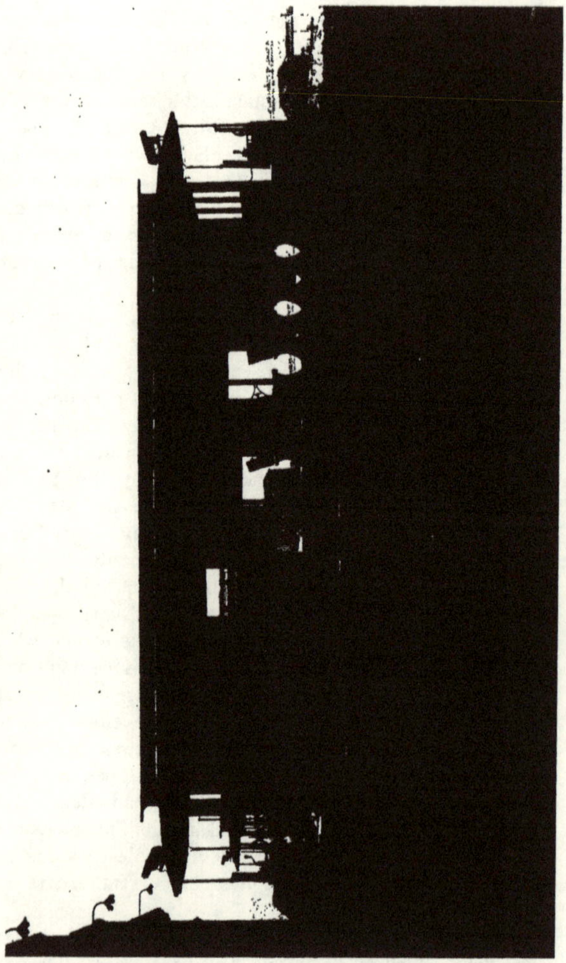

Standard Combination Car

ing plates, which are renewed as often as necessary. The body of the plow which passes through the slot is composed of three pieces of sheet steel, with spaces between them for the flat insulated wire to run down to the contact shoes. The latter are fastened directly on springs, and two flexible conductors run to each shoe to avoid trouble from the breakage of one. The lower part of the plow is entirely of wood, with the exception of the contact shoes, springs and iron for spring support.

The cars of the Metropolitan System are housed in eighteen houses, as shown below. The number of cars shown as operated from each house is the maximum kept ready to be sent out at the rush hour. The number shown as stored is the total capacity of all floors of the house.

	Operated.	*Stored.*
218th St. and Tenth Ave.....	...	150
152d St. and Eighth Ave....	...	24
146th St. and Lenox Ave	220	410
129th St. and Amsterdam Ave..	68	68
129th St. and Third Ave.......	132	297
99th St. and Lexington Ave....	189	264
96th St. and Second Ave.......	190	390
86th St. and Madison Ave......	73	73
65th St. and Third Ave	102	148
54th St. and Tenth Ave......	83	83
53d St. and Ninth Ave	155	155
50th St. and Eighth Ave.......	67	67
50th St. and Seventh Ave......	184	184
50th St. and Sixth Ave........	134	388
42d St. and North River.......	52	65
32d St. and Fourth Ave........	73	73
23d St. and Eleventh Ave....	55	55
14th St. and Avenue B.........	120	184
	1,897	3,078

At all of these houses, except 152d Street and 218th Street, which are used only for the storage of cars out of season, the ordinary routine of inspection and running repairs is carried on. Every car is given a general

Typical Car House

inspection each night. As far as possible this takes place as the car is run into the house. It is first run over the inspection pit and, if found all right, is run back to a storage track; otherwise it is sent to one of the repair pits.

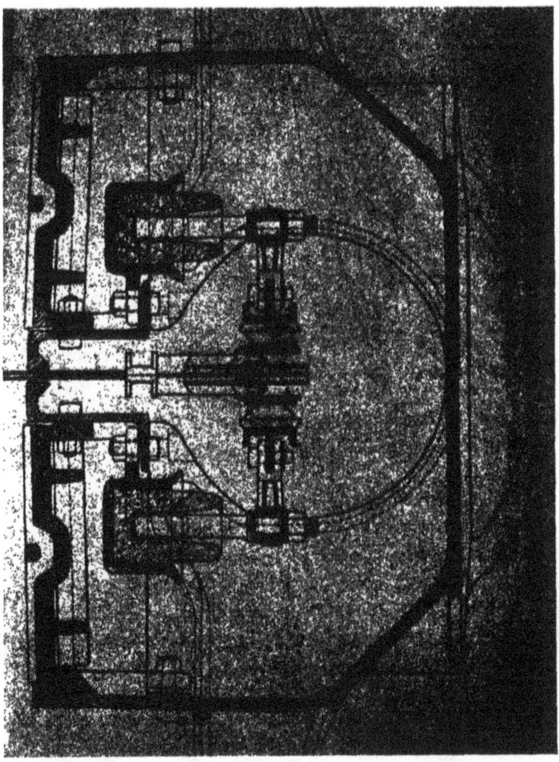

Section of Track showing Pendant Insulation, Plow, etc.

When cars come in very rapidly, it is frequently necessary to set some of them aside to give the men time for inspection. The latter consists in examining the wearing plates, springs and shoes of the plow, washing the plow bar with kerosene and greasing it, examining the motors for loose bolts and hot boxes and the cars for damaged wheels, defective gong or loose truck bolts. An inspection pit is the length of one car, five feet deep in the clear, and its "crew" is three men. No repair work is done there. Every two months cars are taken in for general overhauling. At this time the armature bearings are usually renewed and sometimes the axle bearings.

All overhauling and inspection in car houses is done in the repair pit. Wheels are also renewed from below and car-bodies are never lifted off single trucks. They are somtimes raised up from "maximum traction" trucks. The pits for wheel renewal lie at right angles to the regular repair pit. The track has removable sections and the wheels are taken out by a jack supported on a truck. All repair pits are equipped with hydraulic jacks.

Armatures, fields, plows and controllers requiring repairs are collected from the car houses in a freight car equipped with a crane and sent to the electrical repair shop, where a sufficient force of men is maintained to make all such repairs for the entire system.

The location of the car houses with respect to the operation of the lines shows a considerable variation in practice which may excite comment. They are in most cases legacies from the old horse car lines, and in very few instances has any site been selected solely because of its strategic value from an operating standpoint. Nevertheless, as in the case of the substations, the location of many could hardly be improved on and few are really disadvantageous. Many of these pieces of property are encumbered in such a way that they must be retained. Others are occupied temporarily pending the completion of the program of reconstruction, when they will become unnecessary and may be sold.

At the present time the total passenger equipment of the New York City Railway Company consists of about 3,000 passenger cars. In addition to this there are thirty express cars, ten snow plows, eighty-six sweepers and twenty-five miscellaneous cars. Of these there are only thirty having four-motor equipments.

The cars are geared to run at a schedule speed of about twelve miles per hour, this involving a maximum speed at times as high as sixteen miles per hour, which is perfectly feasible in the upper part of the city. In the lower part of the city the speed frequently falls to eight miles per hour, or even lower.

Owing to the exceedingly heavy service on the longitudinal lines, with the frequent accelerations necessary, the consumption of electric current per car mile is unusually high, two and seven-tenths kilowatt hours per car mile, and demands on the motor equipment severe, with the result that the life of motors, controllers, plows, etc., is comparatively short.

TRANSPORTATION

The car service is, of course, the feature of controlling importance for any street railway, and all the operations of the Metropolitan system are contributory to the regularity and promptness of this service and to the comfort and safety of passengers. As to the immediate agencies for obtaining this regularity, the time table, as used on the Metropolitan system, is merely a schedule of the starting times of the regular "runs" or crews from the terminal. Running times between certains points are given for information, especially when the cars are on long headway, but, as a rule, once out on the road, the headways is governed by the distance from the leader, and, where the vehicle traffic is not seriously congested, depends largely on the efficiency of the crews in avoiding and overcoming delays. In this they are closely watched and checked by uniformed inspectors stationed along the lines, each having a district assigned him, its length varying with the local conditions. When gaps occur in the service they are filled by "extra" cars run out from the nearest car house by the Starter in charge, or by the "switching back," under orders of an inspector, of cars from the opposite track. The transportation service of each line or division is in charge of a General Foreman. He is freed as far as possible from office duties and spends much of his time on the street, becomes very familiar with the variations and needs of the traffic, and, where the time table does not furnish a sufficient supply of cars, arranges for extra service to fill the shortage. He assigns, instructs and disciplines the men who are sent to him from the School of In-

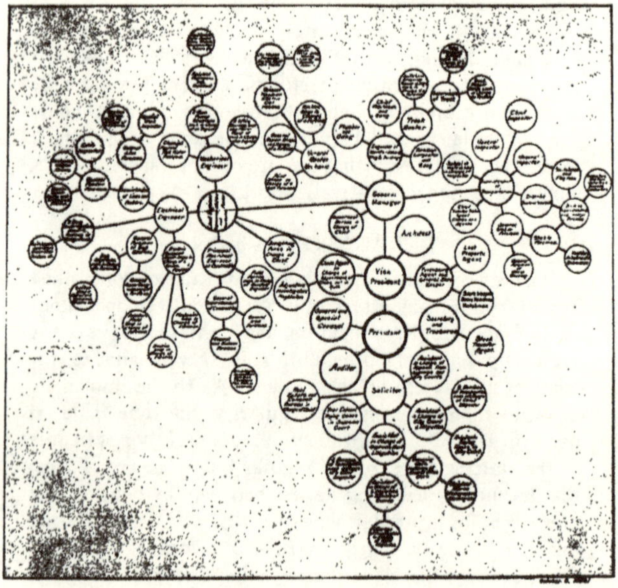

Scheme of Organization of The New York City Railway Co.

struction, and is assisted in these duties by the Inspectors and Starters, the former being especially road men, while the latter are stationed at car houses and terminals to assign the men, fill vacancies and act as despatchers generally.

The system is divided into two districts, eastern and western, each in charge of a superintendent, to whom all the general foremen of the lines in his district report. The special object of this office is to harmonize the service on adjacent lines and at terminals and prevent the irregularities that arise where conflicting authorities come in contact. These superintendents report to the Superintendent of Transportation, who is in general control of the whole service.

The task the Metropolitan Street Railway organization performs is the transportation of upward of a million and a half passengers daily throughout the year. The accompanying chart, showing graphically a day's traffic on one of the important lines, and the following statistics of traffic of an average week day in a moderately busy season will show in a general way how the travel is distributed:

TRAFFIC—METROPOLITAN STREET RAILWAY SYSTEM—DECEMBER 17, 1902.

Date of record, Dec. 17, 1902.		
Total passengers, includings transfers, on system		1,625,127
Passengers by lines—North and South:		
East Side lines	579,092	
West Side lines	644,079	
		1,223,171
Passengers by crosstown lines, East and West.		401,956
Estimated number of passengers North in maximum hour		69,000
Estimated number of passengers South in maximum hour		51,000
Estimated number of passengers East in maximum hour		19,000
Estimated number of passengers West in maximum hour		20,000
Maximum hour on the average day		5-6 P. M.
Estimated percentage of total twenty-four hour traffic handle in five busy hours, 7 to 10 A. M. and 5 to 7 P. M.		35%

The maximum car service on the date referred to

was about 1,600 cars, between 5 and 6 P. M. It must be remarked that a large part of this traffic is moved through narrow streets, densely congested with vehicles of all kinds, under no effective constraint to give the car the right-of-way, and yielding such privilege only

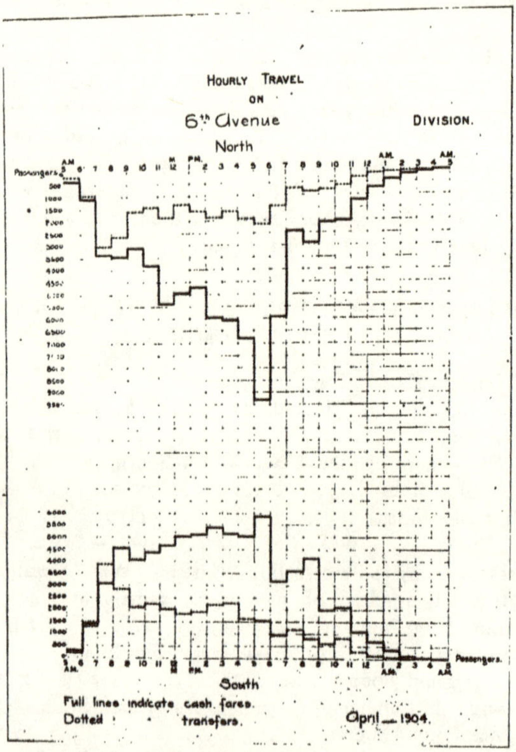

Hourly Traffic Chart

by virtue of the good will of the driver. On the Broadway, Fourth Avenue and some other lines a headway of fifteen to twenty seconds is maintained for several hours in the busy part of the day. Both of these lines are intersected by the busiest lateral thoroughfares in the city and pass in some portion of their route through streets so narrow that a vehicle standing alongside the curb must be passed with caution.

It is inevitable that under such conditions many accidents will occur—especially collisions with vehicles. The wonder is that they are not more numerous. The natural result is a great volume of claims. These, if valid and reasonable, are settled promptly, but as many are excessive, or otherwise unfair, or fraudulent, they pass into litigation, constituting a considerable proportion of the calendars of the higher civil courts. The personal injury claims arising out of street railway accidents in New York since the introduction of mechanical traction have become, both from their volume and from the methods pursued by certain members of the legal profession who have specialized this class of practice, a matter of serious consideration. While the railway companies win the majority of cases that come to trial, the expense of defending so many suits is very great.

As independent companies before consolidation, many of the lines now composing the system traversed meandering and circuitous routes. With the gradual amalgamation that has come about with the change of motive power, these routes have been to a great extent obliterated, and not only does the map present the general aspect of a gridiron, but, disregarding some of the horse lines not yet converted, the "routing" of cars conforms largely to direct north and south and east and west lines. Except for Broadway, whatever, diagonal movement is desired is now obtained by means of transfers.

In addition to the passenger service the Company, under contract with the Metropolitan Express Company, operates about thirty express cars. At the depot at Lexington Avenue and 129th Street, where connection is made with the Union Railway Company, the trolley pole, with which all these cars are equipped, is substituted for the plow, permitting the express cars to reach every part of the system in the Bronx and Westchester county.

With the adoption of cable power on Broadway in 1893, a school of instruction for the "gripmen" (as the drivers of the cable cars were called) was started. During all the changes of power and additions of electric mileage it has kept pace with the requirements of the service, furnishing, since its establishment more than 10,300 trained gripmen and motormen. At times it has been under great pressure, as in 1901, when, in changing the motive power of the cable lines to electricity, it was necessary to qualify their entire force of three hundred and fifty gripmen as motormen, in addition to supplying men to the existing electric lines.

Every applicant for the position of motorman is required to pass a rigorous physical examination, including tests for eyesight, hearing and color-sense. This done, he enters the school, where he becomes thoroughly familiar with the apparatus he is to handle and acquires, amid quiet surroundings, an instinctive knowledge of how to manipulate it under all circumstances before encountering the embarrassing conditions of actual service on the street. Upon certification of the chief instructor as to his ability he is then placed on a line having little street traffic where, under the supervision of an instructor, he operates a car for a week, or ten days, or longer, until he is deemed competent to take charge of a car on a more crowded street. He is then assigned as an extra man to some division on the requisition of the General Foreman. Of the men who pass the physical examination only about forty per cent. finally qualify as first-class men.

One of the primary objects sought by those who planned the consolidation of the roads on the Island of Manhattan was the promotion of travel by an interchange of transfers. As the system grew and mechanical traction was applied to one line after another, the privilege was point by point extended and its use increased enormously, so that, from an issue equivalent to five or six per cent. of the cash fares collected, the percentage rose in 1901 to fifty-six per cent., and the average rate of fare declined from 4.75 cents to 3.16 cents.

At the same time the number of paying passengers had risen from 63,000,000 to 303,000,000, showing how, in conjunction with electric traction, the extension of the privilege had popularized the use of the surface cars. As might be expected, however, gross abuses of the privilege had arisen until a large proportion of passengers were riding to and from their homes daily on a single fare. The cause of most of the abuse was that with the enormous increase in the issue, conductors had no time to punch the date, time and other limitations properly or at all, nor, indeed, to examine the limitations on the tickets presented, and the public becoming familiar with the situation, found that any ticket, however valueless it might be, could be presented on any line with the practical certainty that it would be accepted. The management, in 1901, found itself confronted with the fact that the people of New York could ride over the Metropolitan lines in any direction, and practically for any length of time for a single fare. Up to that time the Company had not asserted its right to place limitations of any kind on the transfer privilege, but it was plainly necessary to do so if the revenue was to be protected. A plan was developed which appeared to meet the demands of the situation and the efficacy of which has been demonstrated by results. Three colors of transfers were adopted, green, red and white, the green tickets to be issued only by conductors upon cars going in a generally northerly direction, the red ticket only by conductors on cars going in a general southerly direction, and the white ticket by conductors on the crosstown lines, good going either north or south. In addition a re-transfer was allowed upon all lines with the exception of the crosstown lines south of 34th Street. In re-transferring on the crosstown lines the transfer ticket was not to be taken up by the conductor, but merely inspected for verification. The printing of the date in large type on each day's tickets avoided the necessity of punching anything but the hour and the point of transfer, and made the ticket worthless after midnight of the date of issue.

This system has continued in use unchanged, and has proved equally satisfactory to the company and to the public. While preventing "circle riding," it enables the passenger to ride from any part of the city to any other part for a single fare and with a minimum of inconvenience in the matter of ticketing.

Europeans are prone to make unfavorable comparisons of the single fare and transfer system with the stage or zone tariff, arguing that the fares of those who desire the transfer are in effect paid by those who do not make use of the privilege and that, as the average fare has been reduced to little above three cents, it would be better for all concerned if the minimum were fixed at three cents, with advancing rates for additional distance or privileges. Whatever there may be in such an argument from a theoretical point of view, there is no possible doubt that in actual practice the transfer system affords the greatest good to the greatest number. If it has been profitable to the Company, its benefits to the public, and especially to people of small means, have been inestimable. For the floating laborer or mechanic, whose place of business is constantly changing, the fact of being able to travel from any part of the borough to any other for a single fare means a great deal. The frequent changes of residence which were formerly necessary to enable the mechanic to live near his job have, since the establishment of the transfer system, almost entirely disappeared, with the resulting introduction of economy and comfort in his home life.

THE EDISON ELECTRIC IL-
LUMINATING COMPANY,
OF BROOKLYN

The Edison Electric Illuminating Company, of Brooklyn

THE Borough of Brooklyn, by reason of its proximity to the Borough of Manhattan, suffers the inevitable loss of prestige which must follow to that which, however great, lies in the shadow of something greater, and the greatness of Brooklyn is seldom realized or adequately understood, not alone by the stranger at its gates, but by the denizen of the smaller, if wealthier borough, on the other side of the East River.

Prior to consolidation in 1896, when it became a part of the greater New York, Brooklyn was in point of population the fourth city in the United States, and an industrial centre of the first magnitude. Since then the growth has been very rapid, so that the Board of Health estimates for 1903 give Brooklyn a population of 1,313,095, as against 1,928,866 for Manhattan. As far as area is concerned, Manhattan is much smaller than Brooklyn, the former covering 21.93 square miles, the latter 77.62 square miles. Within this territory one finds not alone that for which Brooklyn is noted—homes and churches—but miles of shops, wholesale and retail, and hundreds of factories. There are also along the water front of the Borough of Brooklyn two institutions which are specially worthy of mention, namely, the very large and commodious piers and docks of the Bush Company, Limited, and Coney Island, a summer watering place whose name and fame are a household word in the United States.

To this constituency, the Edison Electric Illuminating Company of Brooklyn supplies electrical energy for lighting and power. The first annual report of the Company shows that on January 1, 1890, there was an equivalent of 6,600 lamps of sixteen candle-power connected to the system, current for which was generated by 250 horse-power engines belted to generators on the floor

Dreamland—Coney Island

above. To-day, there is an equivalent of upwards of 800,000 lamps of sixteen candle-power connected to the system, and the load has grown so rapidly that the 30,000 horse-power in high-class engines at the Company's two waterside stations will soon be insufficient to carry it, and two more engines have been ordered and are now in course of construction—one a 5,000 horse-power cross compound Westinghouse, and the other a 10,000 horse-power turbine, the first Parsons turbine to be made in this country by the Allis-Chalmers Company.

The current as generated at these waterside stations is three-phase alternating, twenty-five cycles, and is carried at 6,600 volts to rotary converting stations, of which the Company has fourteen located in different sections of the Borough, from which it is distributed as direct current at 115 and 230 volts. The entire distributing system is meshed so that in the event of disabling of units, the load is automatically transferred and carried by the remaining units, and, as a still further guarantee of continuity of service, there are eight storage batteries, located at substations, with an aggregate capacity of 65,000 ampere hours. The customer is thus certain to be reasonably exempt from interruptions to the service, while the batteries can be economically charged from the surplus energy of a light day load and used to advantage at the peak.

The Company also operates a two-phase alternating sixty cycle overhead system for suburban or sparsely settled sections and for localities not likely to be specially remunerative until more fully developed. These alternating current circuits are changed over to direct as rapidly as is warranted by business returns. It has been the policy of the Company for years, in coöperation with the city authorities, and in line with service developments, to remove from time to time as much overhead construction as is feasible, substituting the more substantial and less obstructive underground system. In 1903, for instance, 193,000 feet of overhead wires were displaced by underground construction. At present the Company has connected to its system 1,200 miles of overhead wiring and about 650 miles of wires in subways.

The Company does the municipal lighting in the Borough of Brooklyn, supplying current to 4,356 arc lamps of nominal 1,200 candle power for street lighting and to 32,086 incandescents of sixteen-candle power for the illumination of public buildings. The total number of

Dreamland, Coney Island—Decorative Tower, 260 feet high

arc lamps attached to the system is over 10,000 and the power load aggregates 15,000 horse power in motors.

A feature of the Company's equipment is the coal handling apparatus recently installed at a cost of $100,-000 at the Bay Ridge power house, for the rapid, automatic and economical conveyance of fuel. The coal is first lifted in the bucket from the boat to the top of a hoisting tower, where it is automatically crushed and weighed. It then slides through a hopper into a car beneath. The cars run on a cable railway and discharge either into a 4,000-ton coal storage pocket or into an extension of this pocket over the boiler room. If the latter, the coal runs by means of gravity into vertical pipes, through which it is delivered on the boiler room floor, the pipes being controlled by valves which permit the coal supply to be turned on or shut off in accordance with the demand of the boilers. The coal stored in the pocket, when required for use, is carried to the top of the boiler room chutes by an endless chain bucket conveyor. The buckets are filled from discharging hoppers provided with stop valves and located at the bottom of the pocket, carried on the chain to a point over the boiler room, automatically discharged and returned empty to the charging hoppers. By this system, one hundred tons an hour can be readily handled from boats alongside to boiler room floor at a cost of two and one-half cents a ton.

The fuel used, in connection with a system of forced draft, is a combination of No. 3 buckwheat and bituminous coals mixed on the boiler room floor or in the furnace in proportions according to the load, one of bituminous to six of anthracite at light load, one to one at maximum. It is not claimed that this is as effective as the pure soft coal (which is impracticable on account of smoke), but it is so within fifteen per cent., while in cost it is thirty per cent. lower. The results from the use of this combination, it may be said, have been extremely satisfactory.

The lighting and power business of the Brooklyn Edison Company, with one notable exception, is, in its general outlines, of the same kind and character as that of similar electric lighting and power companies in the other large cities in this country. It may be said that there are probably in proportion to the amount of current consumed in Brooklyn fewer isolated plants than are to be found in most of the large cities in the United States. As an illustration of this, it may be cited that in Brooklyn there are twenty-one theatres, current for

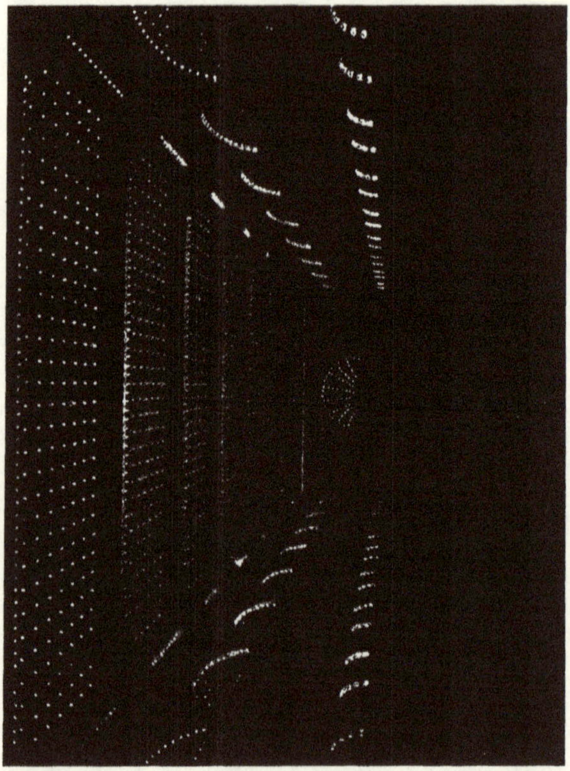

Dreamland, Coney Island—The Ball Room, photographed at night

every one of which is supplied by the Brooklyn Edison Company. Another noteworthy feature of the business here is the large number of electric signs which have been recently installed on the Edison system—about two hundred since January 1, 1903. This effective class of advertising is becoming increasingly popular in Brooklyn. It is profitable business, too, for the company, a larger proportion of the lamps being more continuously in use than is commonly the case with ordinary mercantile or residential lighting.

The Company has also recently taken measures for stimulating and developing the electric automobile business. It has established ten charging stations operated directly by the company and has encouraged the installation of outfits for public use in stables and garages to such purpose that there are now a comparatively large number of them connected to the system.

The exceptional feature of the Brooklyn Company's business is the extent and proportions of its summer load, the peak of which last year was higher on July 4 than on the heaviest day in the previous December, and which this year bids fair to be very much higher than that of any day last winter.

It is unnecessary to point out here that during the summer season, the consumption of current for residential and mercantile lighting is at its lowest ebb and that for the average illuminating company the dog days are encompassed by an all-pervading dullness. The Brooklyn Edison Company is fortunate in that included within its sphere of operation is Coney Island, the greatest summer resort in America, the Mecca of all warm weather visitors to New York and the playground and theatre of amusement to a population within the metropolitan district exceeding four millions. The Island load of the Edison Company begins about the 15th of May, lasts clear through until the 1st of October, and is heaviest in midsummer, when the ordinary run of business is lightest. The development of electric lighting at Coney Island, practically all of which is supplied by the Edison Company, has within the past two years been very striking, due to the advent of great and extensive amusement enterprises, which rely mainly for their power of public attraction upon the brilliant and artistic utilization of myriads of incandescent lamps. The admirable and striking effect of upwards of 100,000 incandescent lamps burning within one enclosure, in ropes, clusters, festoons,

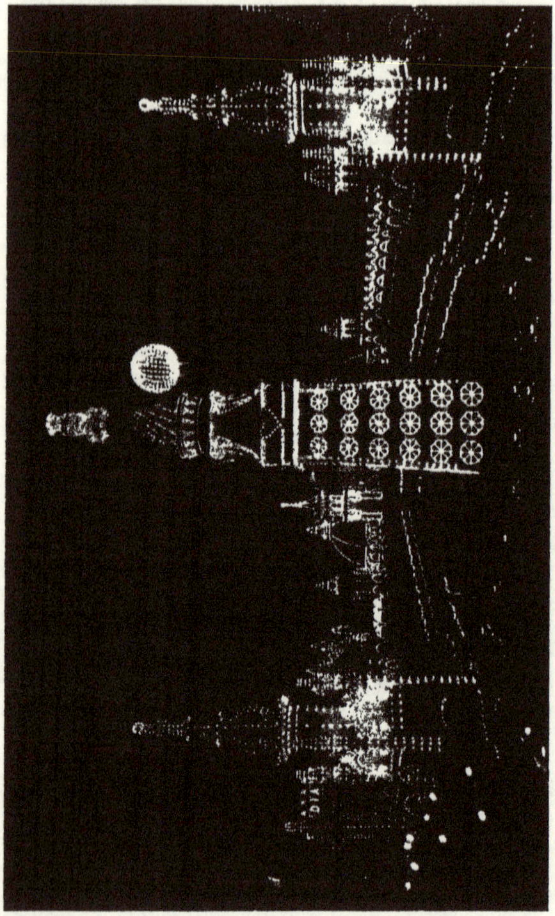

Luna Park, Coney Island

long rows, and illuminating the decorative outlines of entrances, arches, turrets, entire buildings and tall towers, cannot be adequately described, but must be seen to be appreciated. Suffice it to say, however, that the Coney Island lighting of to-day is probably the most lavish and magnificent decorative illumination to be found anywhere in the world.

The most recent of these big Island amusement enterprises is called "Dreamland." It is a combination of vaudeville, spectacular exhibitions and amusement for people of all ages, exteriors and interiors alike resplendent with electric light. Within this one enclosure there are more incandescent lamps than are to be found in many large cities. The decorative tower in the centre alone contains 42,000 lamps, one thousand of them being studded in the ball at the apex. In addition to the incandescent lamps, there is utilized in "Dreamland" about 750 horse power in motors, the entire installation requiring from the Edison Company about 45,000 amperes of current.

Another amusement enterprise of the same character, which is entitled to the credit of priority, and in which the lavish use of incandescent lamps is a notable feature, is called "Luna Park." The Edison service is also extensively employed by a great many other distinctive and interesting amusements and spectacular displays at this resort, individual mention of which is unnecessary.

Most of the lamps in use at Coney Island are of eight-candle power, and this summer (1904) their number will be upwards of 300,000. The effect of such an extensive utilization of current is apparent in the load curves which accompany this article, and coming at a season of the year when ordinary business is lightest, it is an especially welcome and attractive development. Nor must it be supposed that it was accomplished without strenuous effort and large expenditure on the part of the Company. The Company's lines were not extended to Coney Island until 1897, and even then the extension, which involved the construction of six miles of high tension transmission line, a rotary converter station and

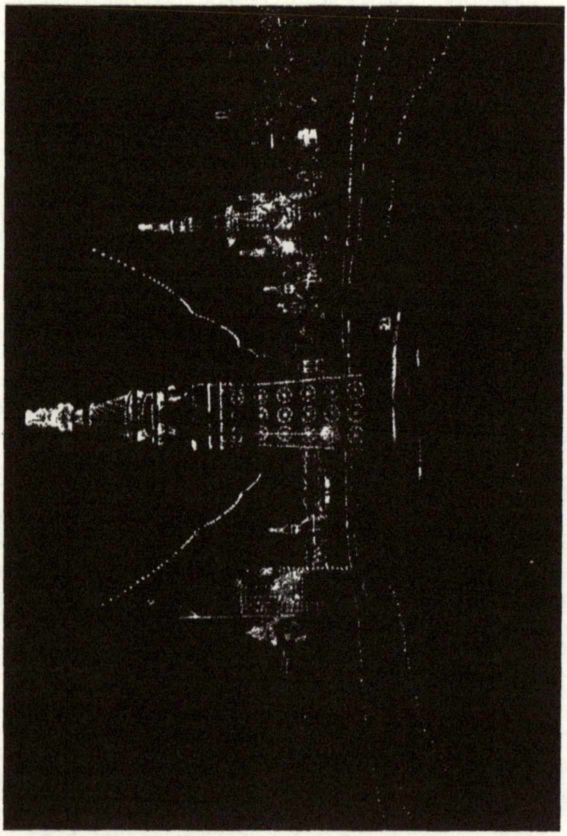

Luna Park, Coney Island

mains and feeders, was opposed by a conservative element in the management of the company, which regarded it as a daring venture, the outcome of which was doubtful. Since then the Company has always stood ready to do its share toward the development of business at Coney Island and at times has been obliged to assume a considerable amount of risk, almost if not quite as great as the cost of the first outlay. As an example, when the plans for lighting Luna Park were first laid

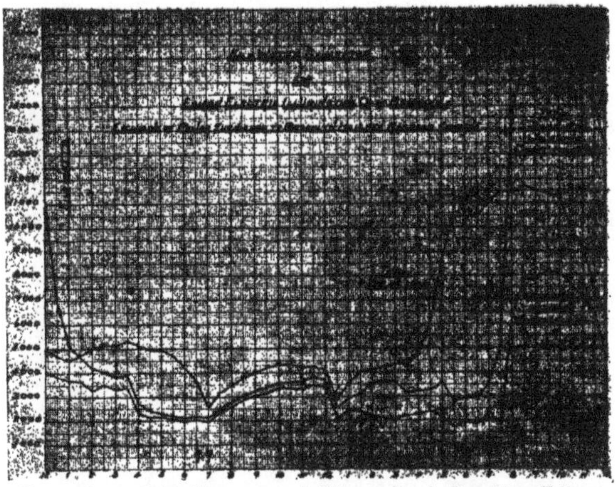

Comparison of the Summer and Winter Load, Brooklyn Edison Co., showing the effect of the Coney Island business

before the officers of the Company, they saw at a glance that the equipment necessary for the substation and the underground low tension feeders and mains would cost many thousands of dollars. With nothing to rely upon as security for their investment except the reputation of the proprietors for business ability and the success which had usually attended all well organized enterprises at Coney Island, they accepted the risk. The result, an extraordinary success for Luna Park, not only vindicated the judgment of the officers of the Edison Company, but stimulated the greater use of electric lighting by many others and was directly responsible for the establishment by a syndicate of capitalists of the "Dreamland" already mentioned. In preparation for "Dreamland" and other Coney Island developments of 1904, six miles of subway were laid, additional rotaries with a capacity of over 5,000 kilowatts installed, a new building erected and many feeders placed, at an expense of over $300,000.

Within the past two years, in conformity with modern methods, the Brooklyn Edison Company has organized a well-equipped advertising department, which works in coöperation with the contract department in obtaining new business. One feature of the work of the advertising bureau is the publication of a monthly illustrated magazine called *The Brooklyn Edison*, which aims to be a medium of communication between the company and the public, chronicling new and interesting facts not only in connection with the operations of the company, but in the wider field of general electrical developments having a bearing upon the lighting and power industry.

THE NEW YORK TERMINUS OF THE PENNSYLVANIA AND THE LONG ISLAND RAILROADS

The New York Terminus of the Pennsylvania and Long Island Railroads

THE solution of the problem of the physical connection of New York City with the great trunk lines terminating on the west bank of the Hudson has long been desired by travelers from the West and South, as well as by the nearby residents of the State of New Jersey. Through railway connection of New England points, by the way of New York city, to the southwest is a matter of scarcely less importance for the convenience of a great section of the country. Furthermore, connection without ferry transfer with the desirable residence tract comprised within the limits of Long Island is of immense importance as an outlet for the ever-increasing population of the city.

The barrier presented by the great rivers surrounding Manhattan Island, as well as the peculiar topography of the city and the high value of its real estate, have presented stupendous obstacles, from both financial and engineering standpoints, to the realization of any practical scheme.

Prior to the successful development of electric traction for heavy railway service the bridging of these river barriers appeared to be the only satisfactory solution of the problem, inasmuch as heavy and congested railway operation through long tunnel lines by means of steam locomotives introduces such serious disadvantages as to nullify any benefits of a central terminus.

The Pennsylvania Railroad Company has had under consideration for many years plans for bringing its terminal from the west over the Hudson River into New York City by means of a bridge, which involved a structure of immense proportions and very costly in itself, and especially so in the city property necessary for its approaches and terminus. This plan, moreover, accomplished only the purpose of a western connection, leaving New England and Long Island points without similar facilities.

Upon the demonstration that electric traction was practicable for heavy train units, the management of the Pennsylvania Railroad again turned its attention to the possibilities of a tunnel entrance. The acquisition by the company of the control of the Long Island Railroad enabled its officers to prepare a well-rounded out scheme for a terminal in the city, embracing local, New Jersey and Long Island business as well as a through connection with New England, thus at once fulfilling all the needs of the situation.

In May, 1902, public announcement was made that the Pennsylvania Railroad Company proposed to enter New York City by tunneling under the North River from the west and the East River from the east and to establish an adequate terminus in the heart of the city,—one which would satisfactorily provide for both through and local business for many years to come.

In the completion of the plans there were many legal obstacles to overcome in securing franchises containing the necessary provisions and involving application to the State and city authorities as well as to the Board of Rapid Transit Railroad Commissioners, which has in charge the development of railway facilities in the city of New York.

The bills enabling the railroad company to enter the city were passed by the Legislature in Albany, April 11 and 14, 1902; a certificate of approval was issued by the Board of Rapid Transit Railroad Commissioners on October 9, 1902, and the franchise was passed by the Board of Aldermen on December 16 and approved by the Mayor on December 22, 1902.

The certificate prescribes the terms of rental, specifies certain uses of the tunnel by the city, and ordains that the work should begin within three months and be completed for operation within five years with a motive power not involving combustion in the tunnels.

Two companies have been incorporated to carry on the work. One of these, the Pennsylvania, New Jersey & New York Railroad Company, will build all of that part of the tunnel and approaches in the State of New Jersey, and extending under the North River to the boundary line between the States of New York and New Jersey. The other, the Pennsylvania, New York & Long Island Railroad Company will construct the tunnel, terminal station and yards, starting from the State line under the North River and extending under the island of Manhattan, the East River and Long Island City. The officers of these companies are Pennsylvania Railroad officers, the president being Mr. A. J. Cassatt, president of the Pennsylvania Railroad Company.

ORGANIZATION

It was evident that this great work would require the highest degree of technical skill to surmount the many engineering difficulties and would demand the closest attention to an infinite amount of detail. It should consequently be undertaken by a well-balanced, efficient and complete organization. The railroad officials, recognizing this fact, endeavored to plan the organization in a thoroughly comprehensive way, aiming at united engineering strength.

The enterprise naturally divides itself into four main departments: the tunnels, the architectural work, the railroad, and the mechanical and electrical engineering. The work as a whole is under the general direction of the management of the Pennsylvania Railroad, which assigned the designing and supervision of construction of the tunnels to a committee of civil engineers; appointed a firm of architects for the superstructure and the architectural work; assumed the task of bringing the road into the tunnels and terminus, and committed the mechanical and electrical engineering to an independent firm of engineers. The general direction of the whole work is assumed by President Cassatt, and its special supervision by Mr. Samuel Rea, fourth vice-president, to whom all departmental bodies report.

Brigadier-General Charles W. Raymond, U. S. Army Engineers, is the chairman of the commission of civil engineers, the other members of which are Mr. Charles M. Jacobs, chief engineer North River division; Mr. Alfred Noble, chief engineer East River division; Mr. William H. Brown, chief engineer Pennsylvania Railroad, in charge of tracks and terminal yards; Mr. George Gibbs, in charge of the mechanical engineering, electric locomotives and traction. Messrs. McKim, Mead & White of New York are the architects of the station, and a comprehensive engineering and construction agreement for the mechanical and electrical engineering has been made with Messrs. Westinghouse, Church, Kerr & Company, who have also been selected as the engineers for the architects. In addition, advisory committees, composed of the operating officers of the road, have been appointed to work out the special problems relating to the required railway facilities and to pass upon the adequacy of the operating features as developed by the labors of the various departmental bodies. Harmony of operation with a proper degree of independence is effected by this organization, which, while it keeps the officers of the Pennsylvania Railroad Company in close touch with the project in all its steps, relieves them of the complexity of work and the multiplicity of detail which are necessarily identified with such a gigantic project. In addition, the task of administration on the part of the Pennsylvania Railroad Company is seen to be much simplified when it is considered that the company has only three bodies with which to deal, namely, the commission, the architects and the engineers, upon whom is placed the whole responsibility and labor of execution.

GENERAL

In general, the scheme adopted comprehends a great terminal station and yard located below the street level in the centre of the shopping district of New York city; the connection of this terminal station by a double-track railway carried in two tunnels westward to a connection with the New York division of the railway on the level stretch of meadows between Jersey City and Newark; connection with the Long Island Railway system at Long Island City by a four-track railway carried in four tunnels; a connection with the Port Morris branch of the New York, New Haven & Hartford Railroad, by means of a new line to be built from Woodside and carried over the East River and Ward's and Randall's islands by a high level bridge; the operation of full size railway trains through these tunnels by means of electric locomotive power; the operation of suburban trains by either locomotive or multiple unit electric power as may best serve the development of the adjoining districts; and the establishment of interchange and terminal service yards both at Long Island City and on the Hackensack meadows, where change from electric to steam locomotive power, and *vice versa,* may be made.

TUNNELS

The tunnel project is the greatest of its kind ever undertaken. The new railroad line will begin at a point on the main line near Harrison, New Jersey, thence over the Hackensack meadows to the yard west of Bergen Hill. At this point it will enter a twin tunnel and drop on a $1\frac{8}{10}$ per cent. grade to the river bank, at which point the river section of the tunnel will begin. These river tunnels are to consist of two tubes driven side by side with 37-foot centers and each twenty-three feet in outside diameter, to contain a single track. The river section will be 4,860 feet in length and the gradients are to be from $1\frac{8}{10}$ to $1\frac{92}{100}$ per cent.

The tubes will be driven by the shield method through the silt composing the river bottom and with the aid of compressed air. Near the east shore of the North River it is expected that rock will be encountered. The general design of these tube tunnels is shown in the figure. It will be noted that they are of heavy flanged cast iron sections, lined with concrete two feet in thickness. The substantial supports for the railway, consisting of steel

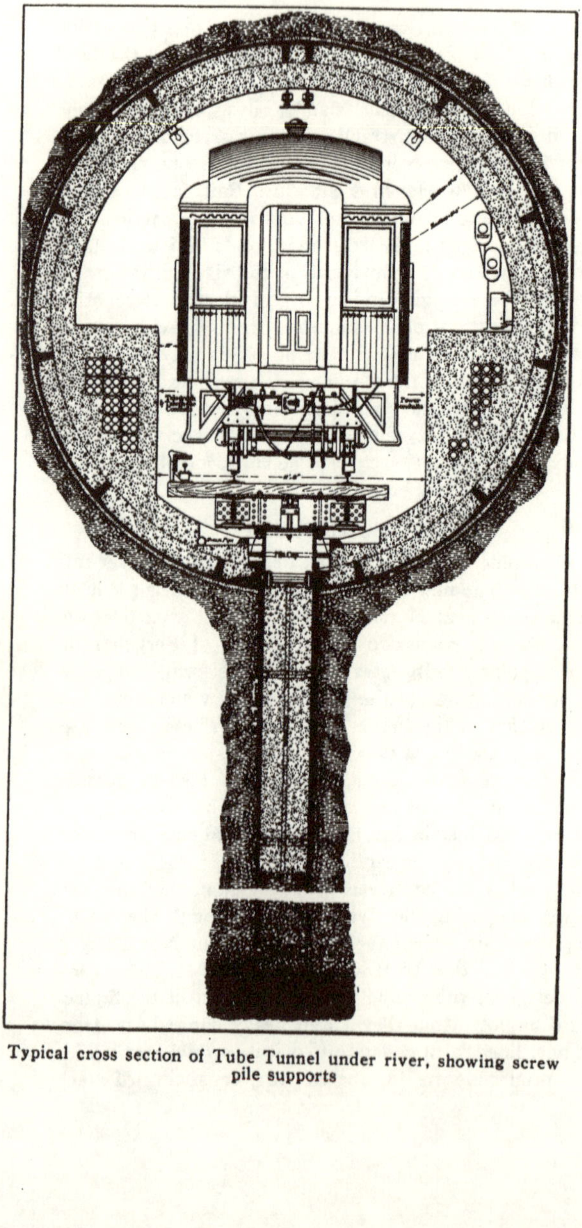

Typical cross section of Tube Tunnel under river, showing screw pile supports screw-pile columns, are also shown, these supports being used in all silt formation. A novel provision, insuring safety in operation, will consist of the concrete side benches extending above the level of the car floor, and which will provide convenient walks as well as housing for the cable-ways, etc.

From the working shaft on the New York side the tunnels are to be continued through Thirty-second street by a twin masonry tunnel to the terminal station yard beginning at Tenth avenue.

The eastern division of the tunnels, giving connection with Long Island, and with New England points through the New York, New Haven & Hartford Railroad, will begin at the Seventh avenue end of the terminal station and proceed easterly across and under the city by means of two double-track twin tunnels, one in Thirty-third and the other in Thirty-second street, to the East River working shafts near First avenue, where again the river sections start. These river sections will be similar to those of the North River, except that the pile foundations are to be omitted and consist in this case of four 23-foot tubes, each 3,970 feet in length, separately driven by the shield method. The gradient for the East River section of the tunnels varies from 1½ per cent. maximum on the New York approach to 1¼ per cent. maximum on the Long Island approach ends.

At the Long Island City shore of the East River the four tunnels emerge into open cuts terminating in an immense interchange yard beginning at Thompson avenue and running eastward for some one and one-half miles.

The mileage of the tunnel system will be as follows:

From west portal in New Jersey to east portal on Long Island, total length	5.6	miles
Single track tunnels	4.72	"
Two-track twin tunnels	4.50	"
Three-track tunnels	0.46	"
Length of tunnel tracks	15.10	"
Length of station tracks	12.85	"
Total tracks between portals of tunnels	27.95	"

It will be noted from the above that the single length of its tunnel line from its portal in New Jersey to its portal in Long Island will be 5.6 miles, and in this distance there are to be somewhat over fifteen miles of single track tunnels.

The actual length of "electric locomotive run," measured from the centre of the interchange yard to the centre of the terminal station in Manhattan, is to be somewhat over four miles for each of the North and East River sections.

TERMINAL STATION

The terminal station site occupies the four city blocks bounded by Ninth avenue, Seventh avenue, Thirty-third and Thirty-first streets, and includes the closing and occupation of Thirty-second street on the surface from Seventh to Ninth avenue; also a large section of the property between Ninth and Tenth avenues on Thirty-second street. The entire tract, measuring about 1,200,000 square feet will be occupied by the terminal station yard at a level of about forty-five feet below the surface. The excavation will involve the removal of some 2,000,000 cubic yards of material and the construction of massive retaining walls surrounding the excavation will require the use of some 50,000 cubic yards of concrete in the walls. The station tracks will be twenty-one in number and the yard will include an area of about twenty acres under four blocks.

The terminal station proper will cover two city blocks and one intersecting street, a total area of about eight acres.

An imposing granite building has been designed by the architects, Messrs. McKim, Mead & White, for the terminal station above the street level. A fine model of this building, showing its relation to the railroad below, is exhibited at the Louisiana Purchase Exposition and is well worth study. This building will be of steel skeleton construction with masonry curtain walls, all to be supported by a system of columns carrying the weight of the foundations to the rock below the track level. The design of this building and its supports is in itself a great engineering problem.

Suitable facilities in the way of entrances and exits will be provided for handling passengers to and from trains. The general waiting room and concourse are to be on an intermediate level between the street and tracks. The former, approximately 100 feet wide by 300 feet long, will be a most imposing feature of the station, its immense vaulted ceiling rising to a height of about 150 feet. The concourse is 100 feet wide and spans the entire track area between the retaining walls for a distance of 510 feet. It is intended to erect over this concourse a groined steel arched roof covered with glass, the roof to be supported on steel columns standing free above the track level in some instances to a height of ninety feet.

It is scarcely necessary to say that the station will embrace all the conveniences required for a modern terminus of the largest size, and is moreover expected to be an ornament to the city and in keeping with the monumental character of this entire enterprise.

The remaining portion of the surface within the terminal yard boundaries will probably be utilized for express buildings and possibly by the postoffice of the United States Government.

As the station tracks are about ten feet below mean high water an elaborate system of under-drainage will be installed, connected to sumps at the service station on the south side of Thirty-first street, at which point the water will be pumped into the city sewers.

South of the terminus a service station is to be provided in which will be installed machinery for lighting, heating and ventilating the terminus and for the operation of the interlocking system. This building will also be used as a substation in connection with the traction system.

CAPACITY OF TERMINAL STATIONS

The design of the track facilities of the New York terminus has been a matter of great difficulty and has engaged the attention of a committee of the operating officers of the railroad and the engineers for many months. It will be readily understood that entrance tracks under the city streets, combined with a rectangular configuration of property, introduce difficulties in design; moreover, the carrying of the supports for the streets and avenues as well as station buildings over portions of the yard property introduces many limitations.

The terminal station has two approach tracks on the west end and four on the east and it is desired to provide for handling trains through these approach tracks at an interval of from two to two and one-half minutes at times. With all these limitations a track layout for the terminus has been worked out, which contains over twelve miles of tracks and provides for all terminal station facilities on a very large scale. The train capacity for the new terminal compared with that of several of the other large terminal stations is as follows:

Station	Total trains in and out for 24 hours	At Maximum Hour
Jersey City	255	23
Broad street, Philadelphia......	553	49
South terminal station, Boston..	840	91
New York terminal station......	1,052	144

CHARACTER OF TRAFFIC

The terminal project is intended for passenger requirements only, and the profile of the tunnel lines, together with the yard arrangements, has been designed for the expeditious and safe handling of the large volume of passenger traffic.

The requirements include, as before stated, both through and local service; the former involving the hauling of the heaviest express train of the modern railway, and the latter both light and heavy excursion and local business.

For the through service the locomotive principle of operation will be adhered to; the steam and electric locomotives being exchanged upon entering the zone of electric operation.

In the excursion and suburban business to nearby towns provision will be made for either the locomotives or the multiple unit control system, the project being planned to give the greatest flexibility in methods of operation so as best to meet the growing demands of increasing population.

In the locomotive service every weight of train from the lightest to the heaviest must be provided for, so that the study of the most economical electric locomotive characteristics has been necessary; moreover, the grades on the tunnel lines are heavy and the speeds on the down grades must not be excessive and on the up grades must not be too low. The attainment of a reasonable schedule under the conditions imposed has, therefore, added to the complexity of the problem. A further consideration and a very important one, is the necessity of utilizing to the fullest the maximum capacity of the tunnel lines at rush hours.

ELECTRICAL CONDITIONS

Electric power for traction will be used throughout in the tunnel and in the terminal station. Direct current at 650 volts will be used at the cars. The power will be generated at three-phase alternating current at 25 cycles and 11,000 volts and converted to direct current at four substations, one in New Jersey, one in Manhattan and two in Long Island.

The power requirements for the tunnel work only are as follows:

	Requirements for Maximum Conditions, Kilowatts
North River section traction	5,200
East River section traction	12,000
Switching in Manhattan terminus	3,200
Total	29,400

The maximum train weight in the North River section will be 770 tons, exclusive of the locomotive; in the East River section 500 tons, the average speed from start to stop averaging from thirty to thirty-seven miles per hour, depending upon the weight of the train. The maximum speed will not exceed fifty-five miles per hour.

In each tunnel conduits are imbedded in the concrete benches. On one side the power conduits are placed, and on the other the telephone and telegraph conduits, all with manholes every 400 feet. The conduit system is a duplicate one, allowing for the complete duplication of the feeder cables in each section of the system. The arrangement of cables between power house and substations is such that any serious delay which might result from an accident to the cables in one tunnel can be avoided by transferring the load in that tunnel to the cables of the adjacent tunnels or from one power house to another, as the case requires.

The number of ducts in each tunnel section will vary to suit the requirements, there being from fifteen to thirty-six power ducts and twenty-eight to forty telephone ducts per tunnel. The total number of duct feet of conduit will approximate 5,000,000 feet. The power ducts are to be intersected every half mile by chambers, where suitable switches will be placed to control the collector rail feeders, and to cross connect the circuits of the adjacent tunnels.

ELECTRIC LOCOMOTIVES

A study of all the conditions has led to the adoption of a uniform class of electric locomotive, having such characteristics that a single locomotive will be powerful enough to haul the lighter trains, which compose, say, 75 per cent. of the number of trains arriving. For heavier trains it is the intention to couple two locomotives and operate them as a single unit upon the multiple unit principle. Two locomotives will handle the heaviest regular train now brought into Jersey City by steam, and this will probably be the heaviest unit advisable to handle over the tunnel grades, having due regard to the existing draw bar attachments.

As the work upon the tunnels and terminals must necessarily extend over some years, it was determined that the question of suitable electric locomotive design should be gone into most systematically, in order that the equipment furnished for regular operation should be as free as possible from defects inseparable from the development of new types,—in other words, there appeared to be an opportunity to eliminate by service experiments many defects prior to the time when it will be necessary to order the equipment for operation. Through the courtesy of the Baltimore & Ohio Railroad Company the tunnel line of that company in Baltimore, which is already equipped for electric traction, was offered to the Pennsylvania Railroad Company as a trial ground for its electric locomotives. It has, therefore, been determined to build at once at Altoona, Pa., two sample locomotives to be placed in service for, say, a year's time on the Baltimore line.

The design of these electric locomotives was thought to be quite as much a mechanical as an electrical problem, and it was, therefore, decided to combine the experience of the officers of the mechanical department of the road with that of the electric manufacturing companies in working it out.

At present writing the electrical and mechanical design has been completed under the supervision of Mr. George Gibbs for two types of electric locomotives, *i. e.*, one having geared and the other gearless motors, and the construction of these locomotives will shortly begin at Altoona.

SCHEDULE

As stated before, it is desired to adopt a running time which, while expeditious, will not result in excessive speeds on the down grades. In the attainment of this object, motor characteristics of the Pennsylvania locomotives have been selected so that the maximum speed possible at the foot of the tunnel grades will not exceed about sixty miles per hour. The attainable schedule for trains of average weight between the middle point of the interchange yard and the middle point of the New York terminus, a distance of about 4.1 miles, will be about as follows, the time including starting and stopping:

North River Division—

East bound .... 6 minutes, 50 seconds
West bound .... 7 minutes, 50 seconds

East River Division—

East bound .... 8 minutes
West bound .... 6½ minutes

INTERCHANGE YARD

The design of the interchange yard has been worked out by a committee of operating officers of the road. The problem involved the expeditious exchanging of steam for electric power and *vice versa,* at the termination of the electric runs. It is obvious that the yard must be capable of exchanging power at certain times up to the limit of capacity of the tunnels on a basis of all locomotive trains. Thus the New Jersey yard must be able to despatch trains on one track at 2½ minutes headway and receive trains on another track at the same interval. The Long Island yard having four main tracks, should theoretically have twice this capacity. The two yard designs have been figured for the above maximum capacity and the character of the proposition determined therefrom, resulting in a very interesting layout. Since, however, the character of the traffic may change considerably in the time required for the completion of the tunnels, resulting in cutting down the proportion of locomotive trains and increasing the multiple unit, the yards have been laid out for building in sections.

POWER GENERATION

In considering the best plan of power generation and distribution, the present loading and future possibilities for extension were taken into account, as well as the requirements of the Long Island Railroad electrification, mentioned under another heading. The safest and most flexible method appeared to be the establishment of two power houses, one located in Long Island near the ferry terminal and another at Weehawken, New Jersey, adjoining the tunnel line on the water front.

These power houses will be designed so as to relay each other and will have capacity for emergency operation of the entire tunnel system and the Long Island Railroad electrification from one power house only.

The construction of the New Jersey power house has not yet begun, but its general plan and equipment will be similar to that of the Long Island City house, the initial portion of which is well under way.

The New Jersey power house will have an ultimate capacity of eight 5,500-Kw. units. The Long Island City power house will have an ultimate capacity of twelve 5,500-Kw. units.

In addition to the above, both power houses will be equipped with separate lighting units to take care of the important requirement of reliable lighting of the tunnel lines and terminal buildings.

For the initial equipment to operate the Atlantic avenue division of the Long Island Railroad the Long Island power house has been placed under construction for the generation of power in the spring of 1905. The building for this work will consist of one-half of the eventual building and will cover a ground area of 200 by 250 feet. This portion will accommodate six 5,500-Kw. Westinghouse-Parsons turbo-generating units, of which three will be immediately installed.

At the present time (September, 1904) the work on this building and power plant is well advanced so that the general characteristics can be inspected.

LONG ISLAND RAILROAD ELECTRIC TRACTION

The electrification plans of this company, which are of immediate interest, are those connected with the operation of the Atlantic avenue division rather than electric operation in connection with the Manhattan terminal project of the Pennsylvania Railroad Company, which will be a later development.

While the Long Island Railroad has its main terminus

at Long Island City, it has another and very important terminus at the junction of Atlantic and Flatbush avenues in Brooklyn.

This terminus is reached by double track railway line from Jamaica to East New York and thence through Atlantic avenue to Flatbush avenue. By agreement with the city, the railroad company has undertaken to remove its tracks from the surface of the street in Atlantic avenue and place them in a subway for a portion of the distance and on an elevated structure for another portion, and to operate passenger trains on this line with a motive power not requiring combustion in the tunnels. This requirement obviously pointed to electric traction, and the early completion of the improvement necessitates the pushing of the electrification vigorously, it being the intention to be ready for operation in the summer of 1905.

The importance of these plans to the public will be seen in studying the local transportation facilities existing in Brooklyn and in connection with the extension of the Rapid Transit subway, now under process of construction, giving track connection at Flatbush avenue with the great system of rapid transit lines of the borough of Manhattan by means of the Flatbush avenue and Fulton street extension and the tunnel under the East River to the Battery. Upon the completion of all these plans a short line will be available for passengers to and from the lower end of Manhattan Island to the eastern district of the borough of Brooklyn and to the entire Long Island Railroad system.

From an inspection of a map showing the location of the Long Island Railroad lines it would appear that the Atlantic avenue electrification may be considered separate from that of lines to be connected with the New York terminus, especially as the early completion of the Atlantic avenue improvement necessitates its electric operation long in advance of the completion of the New York terminal project.

In providing for the early electrification of this route the company has planned for the progressive extension of

Map of the Pennsylvania Terminal Railroad and Western Lines of the Long Island Railroad. Portion to be converted to electricity shown by heavy lines. Connection of Atlantic Ave. line with Interborough Railway System also shown and the lines of the New York, New Haven & Hartford Railroad

this method of operation as fast as connection facilities at the various terminals will warrant.

The initial scheme which is now under consideration may be described as follows:

Electrification of the Atlantic avenue line from Flatbush avenue to Jamaica and thence to Belmont Park, a distance of 14.12 miles.

Electrification of the Rockaway division from Woodhaven Junction on the Atlantic avenue line to Rockaway Park, a distance of 8.53 miles.

Electrification of lines from Jamaica to Metropolitan race track, a distance of 2.6 miles.

The total mileage reduced to single track mileage comprehended in the present electrification plans is about eighty-six miles.

The service contemplated will include all business originating and terminating at Flatbush avenue to and from the Atlantic avenue and the Rockaway division. This will include the regular suburban service between Flatbush avenue and Jamaica and between Flatbush avenue and Rockaway Beach; also excursion movement between these points, which latter movement includes a heavy business from Brooklyn to three race tracks.

The train service provided for will be much heavier than that at present in force for steam service, as it is intended to provide for the maximum capacity over the lines which may be required for the growth of the normal suburban service and for the handling of special excursion service.

The trains will be operated on the multiple unit principle, composed of motor and trailer cars. The trains will vary in length from two to eight cars each, according to the traffic requirements. The train intervals will also vary according to requirements, up to the minimum interval of about 2½ minutes, which is the maximum allowable under the protection of the automatic block signal system.

Current collection will be by means of a protected third rail, laid the standard distance from the track rail, adopted by this road, the Pennsylvania and the New

York Central, namely 27 inches from the gauge line of track to the centre line of third rail and with top of rail 3½ inches above the top of the track rail. This will also allow physical connection to the lines of the Interborough System.

The current will be supplied from the Long Island City power house, from which A. C. current at 11,000 volts potential will be led to substations for transformation and conversion into D. C. current at 650 volts. Substations will be located at the following points:

Substation No. 1—Near Flatbush avenue.
" No. 2—At East New York.
" No. 3—At Woodhaven Junction.
" No. 4—Near Rockaway Junction.
" No. 5—At Hammel.

A portable substation, for race movement, will be located at Belmont Park or Metropolitan race track, as required.

For these substations, now under construction, there is provided an initial equipment, built by the Westinghouse Electric and Manufacturing Company, as follows:

Flatbush Avenue—Three 1,000-Kw. rotary converters, with nine 375-Kw. transformers.

East New York—Three 1,000-Kw. rotary converters, with nine 375-Kw. transformers.

Woodhaven Junction—Three 1,500-Kw. rotary converters, with nine 550-Kw. transformers.

Rockaway Junction—Two 1,000 Kw. rotary converters, with six 375-Kw. transformers.

Hammel—Two 1,000-Kw. rotary converters, with six 375-Kw. transformers and one 3,200-ampere hour storage battery.

The design and installation of the electric traction work of the company has been entrusted to Messrs. Westinghouse, Church, Kerr & Company, who have planned the work throughout in harmony with the Pennsylvania terminal project. The general supervision of the work is under charge of a committee of operating officers of the Long Island Railroad Company, of which

Mr. William H. Baldwin, Jr., president, is chairman and Mr. George Gibbs electrical engineer.

At the present writing, the work of installation is proceeding rapidly, over one thousand men being employed in the field.

THE COMMERCIAL CABLE COMPANY AND THE POSTAL TELEGRAPH COMPANY

The Commercial Cable Company and the Postal Telegraph Company

THE COMMERCIAL CABLE COMPANY

THE COMMERCIAL CABLE COMPANY owns and operates between Europe and America four complete transatlantic cables of an aggregate length of 13,212 nautical miles. It also operates the American ends of two direct and entirely submarine cables between Germany and America by way of the Azores. Two of its own cables and both of the German cables run into and are operated from the Commercial Cable Building. The main operating room of the Commercial Cable Building is in direct connection by means of a specially strung heavy copper wire with the operating room of the Commercial Pacific Company's cable at San Francisco and is also in direct wire communication through the Postal Telegraph system with all the principal cities of the United States and with Montreal in Canada.

The Company accepts at its counters in the Commercial Cable Building messages for transmission by cable to Europe, Asia, Africa, South America and Australasia and for the Hawaiian, Ladrone and Philippine Islands and Asia by way of the Pacific Cable, and for Bermuda and the West Indies by way of the Halifax and Bermuda and direct West India cables.

The Commercial Cable Building, twenty-three stories high, is situated at 20 Broad Street, in the financial centre of New York. On one side it adjoins and has direct communication with the New York Stock Exchange, the entrance to one of the Stock Exchange Building's anterooms being located in the main corridor of the Commercial Cable Building. It was erected to provide a permanent home for the Commercial Cable Company's principal traffic office. The two lower floors of the building are used exclusively for telegraphic purposes, but the growth of the Company's business has been so rapid that the accommodation already seems cramped and more space will have to be provided at no very distant date.

The cables belonging to the "Commercial" system which enter New York are landed at Coney Island, the Cable Hut being situated at a point one thousand feet east of the Oriental Hotel. The ocean cables enter the Hut and are there permanently joined to specially designed cables leading to the Company's main operating room. The route followed by the special cables is across Sheepshead Bay, through the Borough of Brooklyn, over the Brooklyn Bridge and through lower New York to No. 20 Broad Street. Through the Borough of Brooklyn and lower New York these cables are drawn into and through iron pipes buried about two feet below the surface of the streets. On the Brooklyn Bridge they are suspended in porcelain insulators clamped to the framework of the Bridge. This line of cables was completed in 1898 and the work was so thoroughly planned and carried out that the cables have been in uninterrupted and satisfactory operation ever since, at the present moment affording connections for five ocean cables. In the case of the more recent ocean cables, in order to minimize the disturbing influence of the trolley currents as much as possible it has been found beneficial to carry a second core ten miles out to sea, and there connect it to the sheathing of the cable. This provides an earth connection as free as possible from local disturbances.

The operating room of the Broad Street office is situated on the main floor and is separated by a glass screen from a corridor which passes through the building from Broad Street to New Street. A separate table is provided for each cable circuit, the sending and receiving apparatus being arranged on opposite sides. Every circuit in the office is duplexed and provided with switches

Head Office of the Commercial Cable Company
20 Broad Street, New York

and apparatus for sending either automatically or by hand.

In mounting the recorders special measures had to be taken to protect them from possibility of mechanical disturbances, due to the vibration of the building or the jarring of the table. In order to accomplish this, a large wrought iron pipe is fastened by means of a flange to the floor of the office. Inside this pipe is placed another but smaller one. The lower end of this inner pipe rests on a pad of soft india rubber one inch thick which completely fills the lower end of the larger pipe. To prevent lateral motion of the inner pipe a ring of half-inch rubber tubing separates the two pipes at their upper end. The platform which carries the recorder has fastened to its lower side a third pipe which can slide up and down inside the inner of the two pipes mentioned above. This third pipe is used merely to adjust the height of the recorder to any desired level. This method has been found most satisfactory in keeping the recorders free from all vibration.

The Commercial Cable Company has eight branch offices in the city of New York, the circuits to which are supplied with current from storage batteries, of which two sets are provided, so that in case of accident to one set the other is always ready for immediate use. As a further reserve, the motor generators used for charging the storage batteries may be used in place of either of them. These two sets of accumulators are also used to provide current for the motors and automatic sending machines. Provision is also made for using the current either from the lighting plant of the building or from the Edison Company's mains, and as a last resort a primary battery is always kept ready for use. Other sets of accumulators are used to provide a low voltage current for the vibrator circuits, etc. All these accumulators are charged by motor generators driven by current taken from the light and power mains of the building.

Primary batteries composed of Fuller's cells are used as main line batteries on the ocean cables. It is inadvisable to use storage batteries for this purpose, as the acid spray given off by the cells during charging renders it impossible to keep the batteries sufficiently well insulated for cable duplex circuits.

At the inner end of the office is a large double spring-jack switchboard, through which the main cable office is connected with the branch offices and with the extensive distributing system concentrated in the chief operating room of the Postal Telegraph Company at 253 Broadway. At the back of the main switchboard is located a smaller one, to which all the cables are led, Lodge lightning protectors being in every case interposed between the switchboard and the cable. Leads also run from the switchboard to the electrical testing room on the nineteenth floor.

The artificial lines used to duplex the various cable circuits are located in a room in the basement devoted solely to their accommodation. They are composed entirely of Muirhead's inductive resistances, with the necessary condensers for sending and receiving. The artificial lines are enclosed in air-tight cases to maintain an even temperature and to exclude dust and moisture.

The Commercial Cable Company was the first to adopt automatic sending machines and to employ vibrators generally on all the recorders in place of the time honored but unreliable "electrified ink."

The types of automatic transmitters, paper motors and vibrators used in this office are all designed by the Company's chief Electrician, Mr. Charles Cuttriss. The distinguishing feature of Mr. Cuttriss's automatic transmitter is the step by step motion imparted to the punched slip by means of which the latter is kept stationary during the transmission of each signal. The transmitter is not connected directly to the cable circuit, but operates in local circuit a set of electro-magnetic cable keys. It is driven by an electric motor contained in the same case. The speed of the motor is controlled by a simple but very effective governor. The contacts are shunted by incandescent lamps to prevent sparking. This transmitter is simple in construction, and the various working parts are easily accessible for cleaning purposes. It is extremely reliable and satisfactory in the way it performs its work.

The paper motors which are used for drawing the slip off the writing platform of the recorders and across the operator's desk are also very reliable in their operation. The speed of the paper is adjustable within wide limits and is maintained at any required value by means of the governor, which is similar in action to that used on the automatic transmitter.

There are two forms of vibrator used in this office. Both types, however, employ the same form of interruptor by means of which the rate of vibration is adjusted. An electro-magnet causes an armature to oscillate to and fro, carrying with it a vertical glass tube, which is connected by a flexible rubber tube with a small reservoir of mercury. By means of a screw, a piston can be raised or lowered in a cylinder connected with the reservoir, and the height of the mercury in the oscillating tube varied, thus affording a continuous and convenient method of adjusting the rate of oscillation.

In addition to the above there are a few combination instruments which are used solely by this Company. The most important provision of all perhaps is the ample reserve power and facilities for quick response to emergency conditions. These reserve and alternative appliances reduce to a minimum the possibility of delay to traffic and keep the system in a high state of efficiency.

THE POSTAL TELEGRAPH COMPANY

The Postal Telegraph-Cable Company operates landlines in the United States which have their focus in the large, well equipped operating room of the Company in the Postal Telegraph Building at 253 Broadway, opposite City Hall, New York, where they connect with the Atlantic system of the Commercial Cable Company, and, radiating from New York, reach every important place in the United States, making connection with the Commercial Pacific cable at San Francisco, and at Montreal with the extensive system of the Canadian Pacific Railway, with which a close working arrangement is maintained. The system embraces 27,482 miles of pole lines and 200,972 miles of wire.

The following comparison shows the remarkable growth of the Company in the last ten years: On January 1, 1894, the Postal system exchanged business between 6,260 offices, and on January 1, 1904, business was exchanged between 19,977 offices, an increase in ten years of 13,717 offices.

The poles of the trunk lines are large and substantial, mainly of cedar, forty to the mile (1.6 kilometres), and the wires are chiefly of hard drawn copper, a large proportion of which weigh 300 pounds (135 kilograms) to the mile, the balance weighing over 200 pounds per mile, excepting a few specially constructed compound wires having a steel core upon which copper has been electro-deposited, which wires, before the advent of hard drawn copper wire, had been erected between New York and Washington, Buffalo and Pittsburgh, and between New York and St. Louis via Chicago. These wires have a steel core weighing 200 pounds per mile, and two wires between New York and Washington have a copper deposit upon the steel of 300 pounds per mile, and two wires New York to Chicago, a distance of 1,000 miles, and one wire Chicago to St. Louis, a distance of 300 miles, have a copper deposit of over 500 pounds per mile, averaging a resistance of 1.8 ohms per mile; and undoubtedly possessing the highest conductivity and having cost more than any aerial wires ever put up for telegraph communication. The poles vary in length from 30 feet (9 metres), set five feet in the ground, to 65 feet, set eight feet in the ground.

The wires are suspended upon one or more well seasoned pine cross-arms, as may be necessary. Each cross-arm is bolted in a gain cut in the side of the pole, and is also supported in position by a pair of iron braces, the ends of which are brought together and bolted to the pole, forming a V. Where more than one arm exists the next arm is placed parallel at a vertical distance of two feet.

The insulators are double petticoat glass, weighing each 1.125 pounds, supported upon the cross-arm by pins which are placed 1.5 feet apart. Lightning rods of wire are conducted from the earth at the foot of the pole and are attached thereto by staples extending above the top about 4 inches (10 centimetres). Notwithstanding the better insulating qualities of porcelain during wet weather, it is found as satisfactory and more economical to employ the glass insulators, for the reason that the average climatic conditions upon the American Continent are such that long continued rains and fogs are confined to very limited areas, and wires are seldom rendered unworkable by reason of insufficient insulation of the glass during rain storms. It is also found, where new porcelain insulators are put up near railroad tracks, that in a few months the locomotive smoke forms a coating upon the surface of the porcelain which reduces its insulating qualities to that of glass insulators similarly situated.

The standard for minimum insulation upon the trunk lines of this Company is 100 megohms per mile in dry weather.

Joints in the hard drawn copper wires are made by the use of copper sleeves, which are practically two tubes six inches in length placed side by side and brazed together. The two ends of the wires to be joined are placed respectively in these tubes and the sleeves are given three complete twists with a steel wrench. This joint requires no solder and wires so jointed have been found to have as good electrical contact at the end of fifteen years as when the joint was first made.

The hard drawn copper wires employed have a resistance as follows:

300 pounds (135 kgs.) per mile, 3 ohms per mile at 60° Fahr.

208 pounds (93.6 kgs.) per mile, 4.22 ohms per mile at 60° Fahr.

In the principal portions of all the cities of the United States such as New York, Chicago, Philadelphia, St. Louis, Boston, Baltimore and Cleveland, this Company's wires are carried in subways beneath the streets in lead

Head Office of the Postal Telegraph Company
253 Broadway, New York

covered cables, which connect the stations in cities with the interurban aerial lines, and foot up more than 200 miles in length, containing a total of over 10,000 miles of conductors.

The wires are arranged as direct circuits connecting the various commercial centres of the country, and are operated exclusively by the American Morse system, the messages being directly transmitted manually by the sending operator through the medium of the electrical dots and dashes. The receiving operator translates the dots and dashes by sound, copying them in full by the typewriter, which is extensively used, or by pen upon the usual telegraph blanks. These messages are copied by a *facsimile* process, enveloped and delivered by messenger to the addressee.

The operation of the wires by the Morse system is upon the simplex, duplex or quadruplex plan, according to the exigencies of the traffic, the single wires being worked upon what is termed the closed-circuit plan. For instance, a wire between New York and Philadelphia, operated as a way circuit, will have sufficient electromotive force supplied at these stations for the operation of Morse relays having a standard resistance of 150 ohms each. Such a relay with a key is placed not only at the terminal stations, but a set is included in the wire at each intermediate station that may be required to transmit its business over the wire, no current being required at any intermediate point for the operation of the main circuit. Between any two points having business enough to exceed the capacity of more than one single wire, duplex or quadruplex apparatus is used.

In case cities are more than 500 miles and less than 1,000 miles apart, it is usual to employ a set of automatic repeaters upon each circuit, whether single, duplex or quadruplex, with an extra supply of current about midway of the circuit, to obviate the use of high potential transmitting currents at the extreme ends of the circuit. Chemical batteries, which at one time were exclusively employed for furnishing current for the operation of main wires, have, during the last twenty years, been almost entirely replaced by dynamos, except for minor local purposes. These dynamo machines are located at the principal main stations of the Company (such as New York, Chicago, Philadelphia, Boston, Buffalo, Pittsburgh, St. Louis, Washington, Atlanta, New Orleans and San Francisco), in all including about fifty-six stations. At each of these stations there are nine motor dynamos in operation, the motor side being operated by any convenient electric light or power current of suitable voltage, usually the same current which furnishes light and power for the building in which the machines are situated. These dynamo machines furnish for the operation of the wires currents of the following voltages:

40 +	
80 +	80 −
120 +	120 −
200 +	200 −
375 +	375 −

One brush of the armature in each machine is connected to the earth. The forty-volt currents are used for all local purposes and for short branch wires in cities. The higher potentials are used for the operation of the apparatus upon the main wires, 200 volts being used for very long, high resistance single circuits, and also for duplexes, and 375 volts being used exclusively in the operation of the quadruplexes.

Between the various apparatus and the dynamos at each station there are inserted inductanceless resistance coils of at least two ohms to the volt, of suitable capacity to protect the switchboards and other apparatus from any damage by heating as the result of accidental short-circuits.

Beside the transmission of the large volume of public telegrams which are handed in to this Company daily over its counters and through its messenger service, a very extended service is performed by its direct wires between the various stock, cotton and produce exchanges of the large commercial cities, the apparatus being so placed in each case as to bring the members of the various exchanges in direct and instant communication with one another. Also many wires are leased to news associations and private customers, connecting them with their agents in all principal cities.

Direct circuits are worked daily from New York to San Francisco, a distance of 3,250 miles; to New Orleans, a distance of 1,334 miles; to St. Louis, a distance of 1,048 miles; to Atlanta, a distance of 882 miles; to Chicago, a distance of 900 miles, and to many other points. These distances contrast strongly with the usual length of circuits elsewhere.

The extensive use of superior hard drawn copper wires, the resistance of which is less than one-sixth that of iron of the same gauge, the improvement in insulation and in the construction of apparatus, the betterment of poles and fixtures, and the large increase in the use by the receiving operators of typewriting machines since their first introduction twenty years ago, have all contributed to the improvement in the service and a higher traffic capacity of the wires.

The following records of speed in Morse transmission, which a few years ago would have been marvellous, are now an almost daily achievement: Between New York and Chicago, on one side of a quadruplex, nearly 1,000 miles, 1,001 telegrams of thirty words were transmitted between 9 A. M. and 5.30 P. M. Of these, 552 passed from New York to Chicago in eight hours and five minutes. From Chicago to San Francisco, 2,700 miles, one operator sent 526 ordinary thirty-word telegrams between 8 A. M. and 5 P. M., being sixty per hour. A quadruplex between New York and Boston carries 2,000 thirty-word telegrams from 8 A. M. to 5.30 P. M. daily.

The Postal Telegraph-Cable Company occupies spacious offices in all the important cities. The main office, at No. 253 Broadway, New York, is in a building which was erected specially for telegraph purposes, and its conveniences and arrangements for the accommodation of the public and the prompt handling of the Company's large and constantly increasing business, as well as the extensive underground cable system in New York City, fairly illustrate the character and equipment of the Company's offices elsewhere.

The first or ground floor of the main office, containing the receiving and delivery departments of the Company, is beautifully finished in marble, the design of the staircase extending from the first to the second floor being of exceptional merit. The ninth and tenth floors are used for the Postal Telegraph-Cable Company's executive offices. The eleventh floor is occupied by the President and officials of the commercial cables. The twelfth floor is used for the main operating room and is equipped with the latest type of apparatus, with the most convenient arrangement of tables and switchboards.

The operating tables accommodating the instruments requisite for the transmission and reception of telegrams are arranged to hold apparatus necessary for eight single Morse wires, or four duplexed or two quadruplexed wires, seating eight operators, four on each side. These tables are distributed throughout the room in such a manner as to afford ample aisle space for the operators, and to insure a quick distribution and collection of telegrams to and from the various operators.

The whole floor is divided into districts containing four tables to each district, and from a central point in each district a pneumatic tube communicates with the gallery located in the centre of the room over the switchboards. All messages arriving over the wires are transmitted to the central station in the gallery and are there assorted, and distributed through the pneumatic tubes for retransmission over the wires in whatever district such wires are located, unless the messages are to be delivered from 253 Broadway, in which case such messages are transmitted through pneumatic tubes from the gallery to the delivery department, where they are recorded, placed in envelopes and despatched to destination by messenger.

Beside the large number of messages which are being sent and received directly by the operators in this operating room, there is an immense volume of traffic that is being automatically transmitted through the room between the large branch offices in New York (such as the Stock, Produce and Cotton Exchanges) and the main or branch offices in other cities.

The various trunk line wires are brought into New York City from the west under the Hudson River by subaqueous cables, and thence by underground cables under the streets to a terminal room in the basement of the Postal Telegraph Building at 253 Broadway. The wires from the north and east are brought through underground cables from the Harlem River to the same point, and after passing through necessary protective devices, reach the operating room on the twelfth floor, where they are connected directly to switchboards provided with springjacks, which allow the instruments and dynamo currents to be connected with the wires as may be necessary.

The switchboards are connected with the various exchanges and branch offices in the same manner by underground cables.

Each switchboard is arranged to contain fifty line wires, which is the maximum number that it is possible for two chief operators to supervise.

Directly in front of the switchboards are located the automatic repeaters which perform the function of forwarding through from one wire to another, or from branch offices through to a distant city, without the intervention of receiving or sending operators. These repeaters are in the care of repeater chief operators.

The pneumatic tubes, which largely replace messengers in collecting and distributing messages in different parts of the room, are operated upon the reservoir system. A reservoir is kept automatically at a pressure of about two pounds, in the engine room, and pipes are led from this reservoir to both ends of a tube which is normally open, but the air pressure pipe connecting with the tube is kept closed by a valve. Carriers, therefore, can be despatched in either direction through the tube. Shutting a small lid at one end closes the pipe and also the circuit of an electro-magnet which holds the lid closed, and at the same time automatically turns on the air pressure behind the carrier placed in the tube. On the arrival of the carrier at the distant end of the tube, it touches a small lever which causes the local circuit of the holding magnet at the sending end to be released, thus allowing the lid to open and the valve to close the air pipe, shutting off the air pressure. This insures a very rapid service and also an economical one, as the tube is utilized for transmission of carriers in both directions, as no compressed air is consumed except during the time the carriers are actually in transit. At night and during Sunday, when the traffic is small, the expenditure of energy for operating the tubes is decreased, the rule being that it is directly proportioned to the volume of traffic.

The underground cables used by the Company have copper conductors 64 mils. (1.62 mms.) diameter, with a minimum of 98 per cent. purity, insulated with three wrappings of best grade of paper fibre to a diameter of $\frac{5}{32}$ of an inch, thoroughly saturated with insulating compound. The lead sheath, containing 3 per cent. of tin, is $\frac{1}{8}$ of an inch thick, and when the cable has been laid each conductor is required to have an insulation of 300 megohms per mile at 60° F., with 100 volts applied for one minute. The decrease of the dielectric resistance by an increase of temperature is ascertained by a table of coefficients which has been determined by experiment. The diameter of the underground ducts in the subways averages three inches, which will easily admit a cable of this kind containing 125 conductors.

Subaqueous cables are usually composed of seven conductors, each conductor composed of seven No. 21 BWG (.81 mms.) wires of a minimum of 98 per cent. pure copper per Matthiessen's standard, and each conductor insulated to $\frac{9}{32}$ of an inch with the highest grade of Para rubber compound. The dielectric of each conductor is subjected in water after twenty-four hours' immersion to a strain of not less than 2,000 volts, applied for one minute, and to show an insulation resistance of not less than 1,000 megohms per mile per conductor, with 100 volts applied for one minute at a temperature of 75° F. The conductors are laid up spirally in the usual manner, with an outside layer of prepared tape and with proper jute bedding, and are protected by a suitable galvanized iron wire armor.

The aerial cables of the Company are composed of No. 14 B. & S. gauge (64 mils., 1.62 mms. diameter) copper wire, at least 98 per cent. pure, per Matthiessen's standard, insulated with high grade of Para rubber compound to a diameter of $\frac{6}{32}$ of an inch. The conductors are laid up in the usual manner with a marking wire in each layer. The first seven wires are laid up with cushioning strands of soft jute yarn and are wrapped with a cotton tape well saturated with a waterproof insulating compound. Between the outside layer and the final outside covering, a layer of jute yarn $\frac{1}{16}$ of an inch thick is applied spirally and well coated with waterproof insulating compound to resist the ingress of moisture, and over this a serving of well lapped waterproof tape, the whole being protected from the weather by a closely woven coat of cotton yarn, which is also thoroughly saturated with a weather-resisting compound. Each conductor is required to have a minimum insulation of 500 megohms per mile at 75° F. with 100 volts applied for one minute when all the other conductors in the cable are connected to the earth.

The underground paper insulated cables are connected at their terminals to short lengths of a special make of terminal cable, containing an equal number of conductors. The conductors are each 51 mils. in diameter, of pure copper wire, and insulated to a diameter of 116 mils. with a good grade of rubber compound containing not less than 40 per cent. of pure Para rubber, over which is placed a good braid of the best quality of cotton, thoroughly saturated with the highest grade of preservative compound. The braid is of a thickness of 12 mils., the diameter over all being $\frac{9}{64}$ of an inch, and the conductors are laid up as usual with a marking wire in each layer, and after being surrounded with a weather-protecting rubber lined tape of 12 mils. thickness, of wide lap, the whole is enclosed in a lead sheath one-eighth of an inch thick, containing 3 per cent. of tin. The conductors of this terminal cable are connected directly to the conductors of the paper cable in the usual manner, with soldered joints, and after being properly protected by thin insulating braid, the joints are enclosed by a sleeve of lead, which is soldered to the lead sheaths of both the paper cable and the special terminal cable. This prevents the moisture from getting access to the highly absorbent paper insulation and permits the conductors insulated with the rubber compound to have the lead sheath cut away at the indoor terminal and the rubber insulated conductors connected directly to the frames in the terminal rooms of the offices.

The Company maintains an extensive system for the sighting of all incoming steamships and of reporting the anticipated time of their arrival at the dock to customers at any stations reached by its lines and connections. For this service a moderate charge is made. Observers are stationed at Fire Island, the Highlands of Navesink, Sandy Hook, Quarantine and City Island, and information relating to arriving vessels is promptly furnished to steamship agents, owners and others who may desire it.

Direct exchange of messages is also made by the Postal Telegraph-Cable Company with several wireless coast stations, at which wireless communications are sent to, and received from, outgoing or incoming vessels fitted with the wireless apparatus.

Telegraphic transfers of money are made for customers between all important offices in the system, in amounts not exceeding one thousand dollars for one customer in one day between first class offices, and in smaller sums to other offices, which are arranged into four classes according to their commercial importance.

BROOKLYN RAPID TRANSIT COMPANY

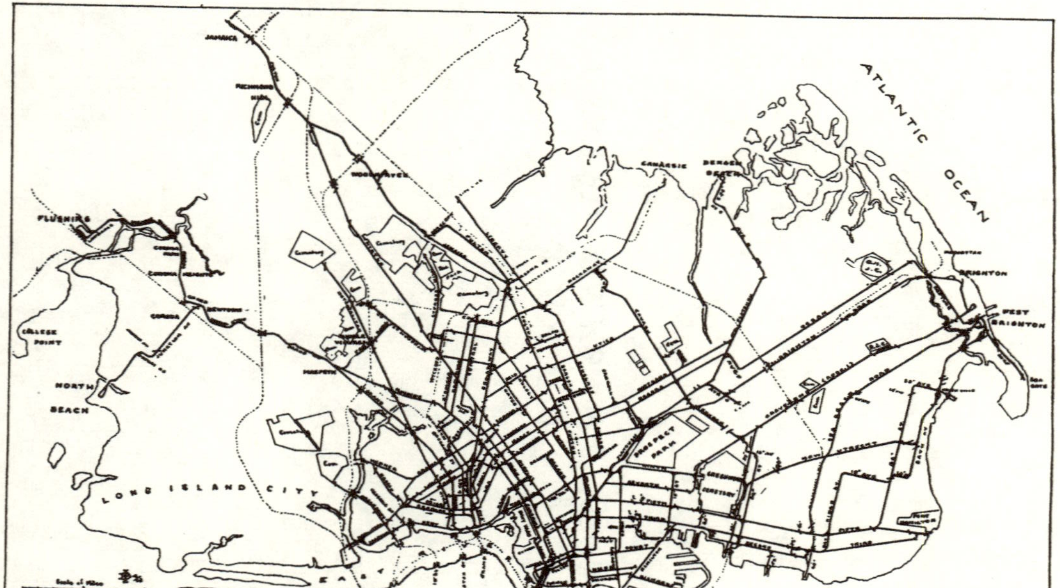

(Brooklyn Rapid Transit System)

Brooklyn Rapid Transit Company

BROOKLYN, the largest in area of the five boroughs of New York city, has unquestionably one of the most complex transportation systems in the world. Practically all of the street car lines in Brooklyn are operated by the Brooklyn Rapid Transit Company, which carries nearly one million passengers daily. Two-thirds of this traffic is carried by the surface cars and one-third by the cars on the elevated railway lines. The large number of passengers carried by the elevated lines, compared with the track mileage and the number of cars running on these lines, emphasizes the value of trains as contrasted with single cars in handling a great volume of traffic. No other terminus on the globe lands and embarks as many passengers as the Manhattan end of the Brooklyn Bridge. One hundred and ten million people are carried across this bridge yearly by cars of the Brooklyn Rapid Transit Company. Every day three hundred and thirty thousand people are carried across the structure. Five cars per minute—two hundred and ninety cars per hour—cross the Brooklyn Bridge on the surface lines alone during the "rush hours," mornings and evenings, when the flood of traffic is at its height. During these hours cars are crossing the bridge every twelve seconds. Hourly 264 elevated cars cross the bridge. More than half of the entire number of passengers carried on the surface and elevated lines across the bridge are engaged in business in Manhattan and live in Brooklyn. The great improvements in transportation facilities and the possibility of reaching a suburban home at a uniform cost of five cents have resulted in the rapid increase in the number of residences in the outlying districts of the Borough of Brooklyn, where ground and house rents are lower than in Manhattan.

The lines of the Brooklyn Rapid Transit system radiate from several points on the Long Island shores of the East River. Owing to the daily morning and evening tide of traffic to and from the Borough of Manhattan, there is congestion at points where the various lines converge.

At the Brooklyn Bridge, as already indicated, the greatest congestion occurs. Congestion, but not to the same extent, occurs at the Broadway ferries in Williamsburg. A large part of this traffic will in future be over the Williamsburg Bridge across the East River, recently completed but not yet ready for surface and elevated railway cars. Other points of congestion are at the ferry landings at Fulton street, at Atlantic avenue, and at similar places.

One of the most crowded street car crossings in the world is to be found at the intersection of Fulton, Willoughby and Adams streets and Boerum place, where many different lines intersect. Here 552 cars per hour cross one point in six directions. The Brooklyn Rapid Transit Company owns and operates 529 miles of track, of which 461 miles are surface and 68 miles elevated tracks. Over these tracks are operated 1,600 surface and 652 elevated cars.

Traffic conditions in Brooklyn are peculiar. Besides acting in part as a dormitory for Manhattan, the Borough of Brooklyn includes within its limits the chief summer resorts of New York City. Of these the most important is Coney Island, which has every variety of amusement from horse racing to sea bathing. The Brighton Beach Hotel, with several hundred rooms and splendid bathing facilities, is owned and managed by the Brooklyn Rapid Transit Company. Other resorts reached by this Company's lines and frequented by hundreds of thousands of people during the season are North Beach, Canarsie, Bergen Beach and Bath Beach. Prospect Park and some 200 smaller parks and picnic grounds scattered throughout the outlying portions of the borough, together with three race tracks and baseball grounds, are also easily reached by trolley or elevated lines.

In the Borough of Brooklyn there are great traffic fluctuations. Weather conditions have a marked effect on the number of excursionists. On pleasant days in the summer the traffic is twenty per cent. greater than on a rainy day. On pleasant Sundays, the traffic is fully 250 per cent. greater than on rainy Sundays. These fluctuations in traffic keep the transportation department continually at work laying out schedules to fit the conditions that arise. Business is mainly regulated by a system of minimum time tables. These are increased when desired by a system of "patches" or sections of the time table which increase the service as needed.

To accommodate the vast throngs which on pleasant days desire transportation to the seashore taxes the carrying capacity of elevated and surface lines of the Brooklyn Rapid Transit Company to the utmost. As many as 375,000 people in one day visit Coney Island and other seashore resorts nearby. To West Brighton, or Coney Island proper, seven surface and five elevated lines are operated. Four surface and one elevated line run to Brighton Beach, and two surface and one elevated line are operated to Manhattan Beach. This gives to the various lines of the Brooklyn Rapid Transit Company a total carrying capacity of some 40,000 people an hour. The elevated structures do not extend to Coney Island. Trains run to that resort over surface tracks. This year the terminal facilities at Coney Island have been remodeled and greatly improved. Thousands of dollars have been spent in improvements at the Culver Terminal.

Because of the rapid increase in population and the still more rapid increase in passenger traffic the power situation in Brooklyn presents difficult problems. For the ten years ending in 1900 Brooklyn increased thirty-nine per cent. in population. But the number of passengers was more than doubled. Power for operating cars of the Brooklyn Rapid Transit system is supplied from seven power stations. Of these the central power station, with 32,000 horse power capacity, is the largest; the smallest is the Montague street cable power house, with 225 horse power capacity. Owing to their having been built at different times and by different companies prior to the consolidation of the transportation interests, the several power stations differ widely in type and size. The electric generating system is partly direct and partly alternating current, with static transformers and rotary converters in six different substations. Alternating current apparatus and steam turbines will be used in all probability in future extensions of the power system. The new central power station is remarkable because of the economical arrangement of machinery as regards space and from the fact that both direct and alternating current is supplied. Six 4,000-horse-power engines supply three-phase, 25-cycle alternating current at 6,600 volts pressure for substations located some distance from the generating stations. Six-hundred-volt direct current is supplied to a territory near the station by two 4,000-horse-power engines. Under this arrangement transformer and converter losses, as well as substation operating costs and interest upon the substation investment, is saved in this territory. In case of any interruption in the alternating current system those lines fed by the direct current generators will not be affected.

In the central power station are thirty-two 650-horse-power tubular boilers made by the Aultman-Taylor Company, arranged in batteries of two and occupying two floors. These boilers have 6,500 square feet of heating surface, are supplied with forced draught apparatus, and the smallest sizes of anthracite coal may be burned. There are no economizers, automatic stokers or superheaters. Provision has been made for subsequent installation of such apparatus if necessary. The engines are of the vertical compound type, directly connected with generators by a drag-crank, and were manufactured by the Allis-Chalmers Company. The cylinders are 42-inch and 86-inch, with 60-inch stroke; a cylinder ratio of 1 to 4.2. The hollow-forged steel shaft is 32 inches in diameter; the valve gear is of the Corliss type; the flywheel 28 feet in diameter and 240,000 pounds in weight.

The engine auxiliaries consist of Worthington jet condensers with air cooler; Worthington circulating air pumps; Worthington rotative dry vacuum pumps, and Wainwright primary heaters. For the reception of the various drips and auxiliary exhausts there are also secondary heaters, and the engines are equipped with steam jackets and re-heaters. These engines are guaranteed to develop at rated load, 4,000 horse-power, upon 12¾ pounds of steam per indicated horse-power, including the steam used for jackets and re-heaters, provided that the steam pressure is maintained at 175 pounds per square inch, that the vacuum is maintained at 26 inches and that the steam does not contain more than three per cent. moisture.

The alternating current generators are each of 2,700 kilowatts capacity. They are of the revolving field type with forty field poles and have a guaranteed efficiency of 96.5 per cent. at rated load with unity power factors. These generators are also guaranteed to carry full load twenty-four hours with a temperature rise of 35° C. above the surrounding atmosphere and to withstand momentary overloads of 100 per cent. The direct current generators are also each of 2,700 kilowatts capacity with a guaranteed efficiency of 95.4 per cent. at rated load. There are three exciters, one steam driven and two induction motor driven, any two of which are sufficient to supply all the field current required. In connection with the exciters a storage battery is used to provide an additional safeguard. There is a large direct current switchboard as well as the necessary alternating current switchboard apparatus, with two sets of high tension bus bars, which may be divided into four separate bus bars by opening oil switches. Solenoids, connected to the storage battery, operate all of the oil switches. All controlling apparatus centres at the operator's board, which is also supplied with a complete set of signal apparatus and a telephone system connecting all parts of the power station and docks.

In planning the station much attention was given to the arrangement of cables. Space was necessary for both high and low tension feeders. Nearly all the cables are lead covered, either enclosed in earthenware conduits or laid in grooves in the walls of the building and covered with stone slabs. To prevent the spread of trouble from one system to the other, alternating and direct current cables are led out from the station in different directions.

Coal can be conveyed from a coal pocket over the boilers from the hoisting tower on the bank of the Gowanus canal or from the storage yard. This yard has a capacity of 100,000 tons and is provided with unloading towers and belt conveyors so that the coal stored in the yard may be transferred to the station at a minimum cost.

Next in size to the central power station of the Brooklyn Rapid Transit Company is that known as the Kent avenue station, near the Brooklyn terminus of the Williamsburg Bridge. This is a direct current station of 16,000 horse-power capacity. Originally the equipment consisted of four General Electric 1,500-kilowatt direct connected units. These at the time of their installation eleven years ago, were the largest railway generators then built. Later on two 1,600-kilowatt Walker generators were installed. These generators are direct-connected to Allis-Corliss cross-compound engines of 2,000 horse-power. Recently a 2,700-kilowatt Westinghouse generator direct connected to a 4,000-horse-power Allis-Chalmers engine has been installed, similar to the two direct current units at the central power station.

Thirty-six 250-horse-power Babcock & Wilcox boilers, hand fired and supplied with engine driven forced draught blowers, form the boiler plant. The older engines have surface condensers, Allis pumps and Corliss valve gear. The station is fully equipped with economizers.

Situated at Fifty-second street and First avenue is the southern power station. It contains six 1,000-horse-power Allis-Corliss cross-compound engines, each belted to two 500-kilowatt General Electric four pole generators. The arrangement of the station is peculiar in that the generators are supported by a centre gallery above and between the two rows of engines with tightening pulleys suspended from the gallery. This permits the use of comparatively short belts. The station was constructed twelve years ago. It marks a period when 500-kilowatt railway generators were the largest built. The boiler equipment in the southern power station consists of sixteen 250-horse-power water tube boilers with Wilkinson automatic stokers. Two firemen attend this plant. Stokers are equipped with forced air blast, increasing the capacity and efficiency and allowing the use of cheaper grades of anthracite coal. In the station are two boosters driven by two Westinghouse 500-horse-power vertical engines.

Automatically dumping cable cars carry coal from the tower at the end of the dock a thousand feet from the station to the coal pocket near the station. The station has a duplicate system of steam piping and larger floor space for each kilowatt unit than many more modern plants. A brick tunnel one thousand feet long by six and one-half feet high by seven feet wide extends from the power station to Second Avenue. Feed wires and ground wires are run from the station on racks arranged along the sides of this tunnel.

Close to the new central power station and near Gowanus canal, at Third avenue and First street, is the Atlantic avenue power station. This contains seven 400-kilowatt generators belted to tandem compound Cooper-Corliss engines; two 800-kilowatt Walker generators direct connected to Allis-Corliss cross-compound engines and two belted booster sets. There are twenty 250-horse-power tubular boilers with Wilkinson stokers. There are jet condensers and primary and secondary heaters.

In the Thirty-ninth street power station at Thirty-ninth street near First avenue are five cross-compound Cooper-Corliss engines, direct connected to Westinghouse generators. Two 1,500-horse-power engines have 1,120-kilowatt generators, two 750-horse-power engines have 660-kilowatt generators and one 750-horse-power engine has a 560-kilowatt booster. These engines have surface condensers, and there are ten 250-horse-power water tube boilers with a coal pocket above.

Two direct connected vertical engines built by the Southwark Foundry & Machine Company and connected to two 400-kilowatt Walker generators are located in the Brooklyn Bridge power station at Washington and Prospect streets. This is the only power station which runs non-condensing and it is operated only during the morning and evening rush hours.

The Montague street cable line, consisting of about one-half mile of track, operating from Wall Street ferry to City Hall in Brooklyn, is operated by the cable power house on State street, where a 225-horse-power tandem compound engine supplies the necessary power. The steep grade near the Wall Street ferry, together with the fact that cars coming down the grade, retaining grip on the cable, assist in pulling cars up the grade, make it desirable from a commercial point of view to continue the operation of this line as a cable road, rather than to substitute electric traction.

There are in the Halsey street substation, at Halsey street and Broadway, five 1,000-kilowatt rotaries, with space for another; in the Tompkins substation, at Tompkins avenue and Fulton street, three 1,000-kilowatt rotaries and one 500-kilowatt, with space for another; in the Essex substation, on Fulton near Essex street, three 1,000-kilowatt, with space for one more; at the Bridge substation, at Washington and Prospect streets, four 1,000-kilowatt rotaries, with space to install two more; in the Coney Island substation on Sheepshead Bay road near Neptune avenue, one 1,000-kilowatt and four 500-kilowatt rotaries; and in the Parkville substation, two 1,000-kilowatt rotaries with room for three others.

Each 1,000-kilowatt rotary has three 375-kilowatt transformers connected in delta with the same number of electrically operated switches on the low tension alternating current side, between the static transformer and the rotary converter. Synchronizing is done on the low tension side of the transformer. While rotaries are supplied with both series and shunt field winding, only the shunt winding is used. The transformers are cooled by air blast supplied by fan blowers driven by direct current motors and are placed on benches which form a conduit for the air supply. The rotaries are brought up to speed for synchronizing by means of a small induction motor, the shaft of the motor being an extension of the rotary converter shaft. Not only are all oil switches operated electrically, but all apparatus is designed and laid out with a view of operating the station with a minimum amount of labor and the greatest freedom from accident. For attending a substation of 5,000-kilowatt capacity there is required but one electrician and an assistant.

Rotary and feeder cables in substations are arranged so as to avoid the possibility of serious trouble from fire or short-circuits. Cables entering the buildings under ground through earthenware ducts extend into the station underneath the switchboard, each duct terminating near the point where its own cable is connected with the board. With this construction there are no groups of cables where trouble may spread from one to the others. High tension cables from the central power station to the substations are usually run underground, the only aerial lines being those which run to Coney Island.

The arrangement of cable lines to the substations is on the "loop" system, so that each substation may be fed from two directions. If by accident all the cables in one line of conduit are destroyed, substations may be fed from the opposite direction. Sufficient cables are provided for each substation to prevent the interruption of service by the disabling of a number of cables.

Storage batteries with a discharge capacity of 2,000 amperes for one hour, or 1,000 amperes for three hours, are in the substations at the Brooklyn Bridge and at Essex street. These help to take up the fluctuations caused by the elevated train service; they are also used during "rush hours" to supplement the generating capacity of the system. These batteries were furnished by the Electric Storage Battery Company.

The line equipment of the Brooklyn Rapid Transit Company follows standard American practice. Tubular iron poles are used. Overhead feed wires are 500,000 circular mils of stranded copper with weatherproof insulation; the cross-arms and pins are of iron; micanite insulators are used; the trolley wire is usually 00 hard drawn round wire suspended from steel span wires. "Figure eight" trolley wire and General Electric grooved wire has also been used in some parts of the line with success.

There are between 700 and 800 miles of overhead feeders, most of which is 500,000 circular mils capacity and a small portion of which is aluminum. There are thirteen miles of 1,000,000 circular mils feed wire. Heavy pole lines have been avoided by the use of elevated structures which follow the lines of greatest traffic. From the power stations to the nearest elevated structures wires are run through underground conduits and over forty miles of lead covered cable is used for this purpose.

Delays from broken trolley wire are rare. Emergency crews cover the entire system and repairs are quickly made. At the New York end of the Brooklyn Bridge, where there are many loops, the use of trolley wire has been abandoned and inverted wooden troughs lined with steel plate are used instead. Overhead return wires are used to a small extent for the return circuit. Particular attention is given to perfecting the underground return circuit, and the rails, usually of the heavy 9-inch girder type, weighing 100 pounds to the yard, are well bonded. Sections near power stations are supplemented by overhead return wires and underground wires. Bare copper cable of 500,000 circular mils capacity is used at one station. Worn out rails laid in trenches and bonded are used for return circuit at two power stations. The elevated structures afford an excellent return circuit, being bonded at all expansion joints, and the carrying capacity of the heaviest structure is equal to forty-five 500,000 circular mils cables. The bonds are usually of the compressed type.

The Brooklyn Rapid Transit Company uses heavy 9-inch girder rails from 90 to 100 pounds per yard on the surface road. Grooved rails are also used. On the Brooklyn Bridge surface lines a heavy flat rail of rectangular cross-section and of special design is used. Not originally intended for car service, no provision was made for laying girder rails on the bridge. These rails have diagonal joints to withstand the heavy bridge traffic and are sixty feet in length.

Different types of permanent rail joints have been tried in Brooklyn within the past few years where track is laid in pavement. This year the Brooklyn Rapid Transit Company is welding over 11,000 joints by the Lorain Steel Company's electric process. This not only secures an absolutely rigid and perfect track, but gives such good conductivity that 100 per cent. of the carrying capacity of the rail is guaranteed.

The third rail used on the elevated lines is second-hand running rail. It weighs from 55 to 65 pounds per yard and is supported upon vitrified granite insulators with iron pedestals for fastening to the ties, malleable iron caps with clips for holding fast on the third rail and a rim or flange for shedding water. The third rail is at the side of the track, 20½ inches from the running rail. Its top is 6¾ inches above that rail. At all grades the third rail is anchored to prevent it from sliding or creeping. This third rail is divided into sections, each several miles in length. Steel brushes on the cars clear the rail from ice.

Elevated cars are operated by the multiple unit system. Five car trains are run during the morning and evening rush hours; two and three car trains at other times. Three motors and two trailers form the five-car trains; two motors and one trailer the three-car trains. The standard style of car is known as the "semi-convertible," a sash pocketing in the side wall with as low a window sill as a comfortable longitudinal seat will permit.

Arc headlights are used on elevated and surface cars. The maximum speed of surface cars is twenty-eight miles per hour and of the elevated forty-two miles per hour. Expresses run from Park Row, Manhattan, to Coney Island, a distance of 11.61 miles, in thirty-four minutes. Besides carrying passengers the Brooklyn Rapid Transit carries the United States mails in special cars between the Brooklyn Post Office and its sub post-office stations.

Cars are also run for the express companies. Brooklyn's ashes are hauled away on thirty gondola cars. Each car has four removal steel bins. Ashes are collected at thirteen ash receiving stations. Electric cranes lift, lower and dump the bins.

Cars are repaired and maintained in seventeen (17) repair shops scattered at the most convenient points all over the system. Elevated cars and surface cars, owing to the difference in construction, are repaired in different shops.

The best surface car shop is located at Fifty-second Street and Second Avenue, and is equipped with the most modern tools, machinery and appliances for the care of surface cars and has a capacity to handle 100 surface cars at once.

The best elevated repair shop is located at Thirty-ninth Street between Second and Third Avenues. Special care was used in the design of the interior of this shop to facilitate the economical repair of elevated cars. It is commodious and well lighted. It is also supplied with the most modern tools and machinery, and has a capacity of handling seventy-five 50-foot elevated cars at one time.

The Brooklyn Rapid Transit Company has its own printing office, where all of its transfers are printed. Nearly half a million transfers are printed daily. Ninety-four varieties of transfers are used. Ten thousand motormen and conductors are employed on the B. R. T. lines during the summer months. In the winter the number employed is about seven thousand. All employees are measured, weighed and examined physically. Eyesight is specially tested. Men are trained as motormen on special cars by instructors. Conductors receive their training by riding on passenger cars with the regular conductor. The wages of conductors and motormen is rated according to the length of service. Increases are given at the expiration of one, three and five years of service. This tends to discourage strikes and is an incentive for the men to continue in the company's service.

The Brown system of discipline with modifications is used with satisfactory results.

Nearly all of the employees of the company are members of the Brooklyn Rapid Transit Mutual Benefit Association. This organization is fostered and encouraged by the company. It is both a benefit and a social organization. Members pay dues of fifty cents a month. If they become disabled or are ill for more than seven days they receive $1 per day and free medical attendance. In case of death $100 is paid by the association to the family. Picnics and other outings to the seaside resorts for the benefit of the employees and their families are held during the summer season. Well equipped club houses with bowling alleys, billiard tables, reading rooms, libraries and gymnasiums are maintained for the use of members of the Association. Small club rooms are located at almost every depot. In these places employees off duty and all extra men find recreation.

WILLIAMSBURG POWER STATION

To provide for the future growth of the system a new station has been commenced adjoining the eastern power station, to be known as the Williamsburg station. This station will be by far the largest and most important power plant of the company and when completed will have a capacity one-third greater than the combined power plants now in use.

The plans provide for an ultimate installation of 66,000 Kw., to be divided between twelve 5,500-Kw. turbo-generator units, of which two have been ordered, one from the Westinghouse-Church-Kerr Company and the other from the Allis-Chalmers Company.

The building will be 257 by 209 feet and will be divided into a boiler house 125 feet 6 inches wide and two stories high, and an engine room 83 feet 6 inches wide and one story high. At one side of the engine room there will be galleries twenty feet wide extending the whole length and across one end of the building for the accommodation of the switchboard apparatus.

The building will rest upon pile and concrete founda-

PROPOSED · POWER · STATION ·
BROOKLYN · HEIGHTS · RAILROAD · CO ·

tions, the piles being uniformly distributed under the boiler room and concentrated in the engine room to sustain the weight of the turbo units.

Central tunnels ten feet in diameter extending nearly the full length of the building will provide for the intake and discharge of condensing water. The intake tunnel is to be located directly beneath the discharge tunnel, but their connection with the river is arranged so as to be as far apart as possible.

The boiler house will be built to accommodate seventy-two 650-H. P. boilers. A large coal pocket of 18,000 tons capacity will be supported from the steel framework of the building.

The 5,500-Kw. dynamos will be so wound as to be capable of connection for either 6,600 or 11,000-volt potential, it being the intention of the company to operate all their alternating current power circuits at the higher potential.

Additional substations are to be constructed in the Bushwick and New Utrecht districts. Subways will connect all the substations with the new station and high tension cables will be installed so that any substation can be supplied from either the Central or Williamsburg stations.

ELECTRICAL EQUIPMENT OF INTERBOROUGH RAPID TRANSIT COMPANY

Electrical Equipment of Interborough Rapid Transit Company: Manhattan Railway Division

ALL of the elevated railways now in operation on Manhattan Island, and in the Borough of the Bronx, were leased in April, 1903, to the Interborough Rapid Transit Company, and are now known as the Manhattan Railway Division of that Company.

The rolling stock required is over 1,500 cars, and, during the hours of heavy traffic, 1,332 of these cars are in service on the road. About 850,000 passengers are carried each day. The heaviest passenger load in the history of the road was on April 4, 1904, when 1,063,000 passengers were carried.

The trains were formerly hauled by steam locomotives, which pulled four and five-car trains at a speed of from ten to eleven miles per hour during the hours of heavy traffic.

In order to increase the carrying capacity of the road, a change in motive power was required which would enable longer trains to be hauled at greater speed. The experience of the electrically operated elevated roads in Chicago, the first of which was put in service at the World's Fair in that city during 1893, demonstrated the advantages of electricity as a motive power for elevated trains and, in February, 1899, the stockholders of the Manhattan Railway voted to change the system of motive power from steam to electricity and issued $18,000,000 in additional stock to pay for this improvement.

The first regular electric train on the Manhattan Elevated Railway System was put into service in January, 1902. A year later eighty per cent. of the entire system was in electric operation, and the last steam train was taken off on June 24, 1903.

The substitution of electricity has enabled an increase of thirty-three per cent. to be made in the carrying capacity of the road, as indicated by the actual increase in car mileage. The electric trains at rush hours are now composed of six cars each and make a speed of thirteen and one-half miles per hour.

74th Street Power Station, Manhattan Division

The passenger traffic during 1903, the first year of entire electric operation, was thirty per cent. more than the last year of steam operation, 1901. The elimination of the smoke, steam and cinders, incidental to steam locomotive operation, is universally appreciated, but the final, and conclusive, advantage of electricity as a motive power is that the cost of electric operation during the year 1903 was less than forty-five per cent. of the gross receipts, while the cost of steam operation in 1901 was over fifty-five per cent. of the gross receipts.

The power for the operation of all trains on the Manhattan Railway Division is generated at one power station located near the centre of the system on the East River, between 74th and 75th Streets. In this station, three-phase alternators, driven by compound condensing engines, deliver power at 11,000 volts potential to the main station bus bars. The alternators are connected to the bus bars through oil switches which control the 11,000-volt circuit. The power is then distributed to substations, located along the lines of the road at approximately equidistant points. Three-conductor cables, placed underground in vitrified clay ducts, convey power from the main station to these substations.

In the substations, the potential is first reduced from 11,000 to 390 volts, by means of step-down transformers, then it passes through rotary converters, which supply direct current at a potential of 625 volts to the third rail for the operation of the trains.

POWER STATION

The power station has a distinctively massive and symmetrical appearance, for the exterior walls, relieved by high arched windows, are carried around the entire building at a uniform height, thus preserving the unity of design. Below the windows, which are about twenty-five feet above the ground, the walls are faced with rough finished granite, and above this line moulded bricks are used. The roof is covered with red tile, with continuous monitor windows above engine and boiler rooms.

The building is 204 feet wide and extends 395 feet along 74th Street and 413 feet along 75th Street. Between the east wall and the dock, a space of eighty-five feet has been reserved by the city for the future construction of Exterior Street. The property purchased for the power station extends back from the west wall a sufficient distance so that the building may be extended to contain additional power equipment.

The foundations rest on bed rock, in which much excavation was required in order to arrive at a uniform level.

The building is of brick and steel fireproof construction throughout. Combustible material has been avoided even in the window frames and sashes, which are of cast iron, and all doors and wooden office partitions are sheathed either with copper or galvanized iron.

A brick partition wall divides the engine room, which

Interior of Engine Room 74th Street Power Station

is ninety-three feet six inches wide, from the boiler room, which is one hundred and four feet two inches wide.

The four stacks are each seventeen feet inside diameter and 278 feet high above the basement floor, and 267 feet above the grates of the lower boilers. The base of the stack is octagonal in shape to a point above the flue opening of the upper tier of boilers seventy-three feet from the basement floor. The walls of the base are five feet thick, with a lining of hollow fire brick supported on corbeled shelves ten feet apart to allow for expansion.

The circular shaft is built of Custodis brick, which are perforated with one-inch square holes. These holes serve to reduce the weight of the brick and also provide dead air spaces, which decrease the amount of heat carried through the walls of the stack.

The power generating machinery has been arranged in eight distinct units, each unit consisting of one engine and condenser, four batteries of boilers, and one boiler feed pump. This arrangement lends itself most favorably to the construction of the building, as it allows uniform spacing and duplication of columns and beams. It also simplifies the piping, which is identical for the eight units, and provides, between each engine and its corresponding boilers, the most direct route for the steam.

The prime movers are eight Allis-Chalmers engines, each of 8,000 horse-power. Each of these engines has a straight shaft supported by two bearings with crank discs at each end and with the rotating element of the alternator located centrally. A high pressure cylinder, placed horizontally, and a low pressure cylinder, placed vertically, is attached to each crank disc, the connecting rods being attached side by side to one crank pin. The crank pin at one end is placed 135 degrees ahead of the crank pin at the other end of the shaft. During each revolution the shaft receives two impulses from each of the four cylinders, thus producing a very uniform rotative effect.

The cylinders are forty-four and eighty-eight inches in diameter, respectively, with five-foot stroke and the normal speed is seventy-five revolutions a minute.

Valves are of the Reynolds-Corliss automatic type and separate eccentrics are used for operating the valves on the high and low pressure cylinders. A ball governor, with an oil operated relay, controls the point of cut-off of the high and low pressure cylinders. The relay consists of a small cylinder with a piston connected to the engine valve gear.

The ball governor is arranged so that its movement opens or closes valves which admit oil under pressure to the relay cylinder. The relay, therefore, serves to make the governing of the engine more sensitive, as the work of moving the engine valve gears is done by the oil piston. The variation in speed, when running in regular service, is guaranteed not to exceed three-fifths of one degree in one revolution.

The steam consumption, when developing 8,000 Brake Horse Power with 150 pounds steam, twenty-six inches vacuum, at seventy-five revolutions a minute, is guaranteed not to exceed thirteen pounds steam per indicated horse-power per hour.

Eight barometric type jet condensers are installed with the condensing cones placed as near as possible to the discharge opening of each of the low pressure cylinders. Between the top of the cone and the cylinder opening, a special "Tee" is placed to allow for the atmospheric exhaust connection. The bottom of the cone is about thirty-four feet above extreme high water, and the discharge tube is carried to the bottom of the discharge tunnel, so that the end of the pipe is always submerged.

The condensing water is supplied to each of the 8,000-H.P. engines by a centrifugal circulating pump direct-connected to a single cylinder engine, the exhaust of which is piped to the receiver of the main engine.

In order to provide against a failure of any one of the centrifugal pumps, a duplex steam-driven pump, having a capacity of 7,500 gallons per minute, is installed and connected to a twenty-four-inch pipe running the entire length of the building, from which connections are made to the injection pipes of each condenser. The vacuum is equalized in the condensing chambers by the use of an equalizing pipe, twelve inches in diameter, which connects the exhaust pipes above the condensers.

Dry air pumps are not required, as a high degree of vacuum is obtained without their use. A motor-driven vacuum pump has, however, been installed with pipes connecting to the discharge of each of the centrifugal pumps, which serves to prime the pumps.

The spray nozzle in the condensing chamber consists of an umbrella-shaped casting, through the center of which the injection water passes. A cone-shaped casting, supported in the centre of the injection pipe on a vertical spindle, causes the water to be distributed uniformly over the umbrella-shaped casting. The advantage of this form of spray nozzle is that there are no small openings to become clogged with foreign matter held in suspension in the injection water, and anything that can pass through the four-inch mesh screen will safely pass through the condenser.

The cost of maintenance on this type of condenser is extremely low, as there are no valve seats, springs or tubes to get out of order. The degree of vacuum produced averages twenty-eight inches, about two inches less than the height of the barometer taken simultaneously.

The water for condensing the steam is taken from the river through a tunnel built below the basement floor of the engine room along the 74th Street side. This tunnel is rectangular in section, 8 ft. 6 in. wide and 14 ft. 3 in deep. Parallel to this tunnel is a smaller tunnel, five feet wide and of the same depth, which carries away the discharge water from the condensers. At the river ends, these tunnels are separated eighty-five feet to prevent the warm water from the discharge tunnel returning through the intake tunnel. At the mouth of the intake tunnel, a set of double screens, with four-inch mesh, is built to prevent the passage of floating material into the tunnel. It is necessary to raise these screens at least once each day to remove the quantity of refuse which clings to the gratings.

The boiler equipment consists of sixty-four Babcock & Wilcox horizontal water tube boilers, each rated at 520 boiler horse-power at thirty pounds steam per H.P. per hour. Eight additional boilers of 600 B.H.P. each were recently installed and will be used for the operation of the new 5,500-Kw. turbo-generator. The boilers are placed on two floors with centre firing aisles which run longitudinally with the building. They are built in batteries of two and are supported on the columns of the building independent of the brick walls and floors. Eight boilers, four on each floor, occupy the space opposite one engine unit and normally supply steam for this unit.

Each boiler has 5,200 square feet of heating surface, and the mechanical stoker has ninety-four square feet of grate area. The normal steam pressure carried at the boilers is 165 pounds above atmosphere, but the boilers are built to safely carry 200 pounds.

The actual evaporation of the boilers with coal having a heat value of 14,000 British thermal units per pound, and with feed water at an average temperature of 188° F., averages 9.18 pounds of water per pound of coal.

The Roney mechanical stokers, with which the boilers are fitted, are capable of burning either anthracite or bituminous coal, or a mixture of the two. The coal is delivered to the stokers through iron pipes from the coal bunkers at the top of the building and requires no manual handling.

An economizer, for heating the feed water, is provided for each set of two batteries of boilers in the original installation, making sixteen economizers in all. Each economizer is placed in the rear of its set of boilers, with a smoke flue between the inner wall of the economizer and the rear wall of the boilers. This flue is provided with dampers, so that the flue gases may be led through the economizer or passed directly to the stack.

For the first three months in the year 1904, the average temperature of the water at the city water main was 34.8° F.; the temperature at which it left the surge tank was 71.7° F., and the temperature on leaving the economizers was 182.3° F., showing an increase of 110.6° F., due to the use of the economizers.

Sixteen blowers are installed in the boiler room for

the purpose of furnishing forced draft in case anthracite coal is used. Each blower has a capacity of 57,000 cubic feet of air at a pressure of one ounce per square inch, and is driven by a direct connected twenty-five horse-power induction motor.

Eight pumps, each having a capacity of 360 gallons per minute, are installed in the centre aisle of the boiler room basement for supplying water to the boilers. Each pump consists of three single acting cylinders driven by gearing from a sixty-five horse-power, 500 volt, shunt-wound, motor.

The steam piping is arranged in eight sections, each section connecting one engine with four batteries of boilers, two on the upper and two on the lower floor. The two batteries on each floor are connected to a short header, eighteen inches in diameter, and these headers are connected between adjacent sections by fourteen-inch equalizing pipes, bent at large radius.

For each engine, a steam reservoir, thirty-six inches in diameter and twenty-four feet long, is provided. The expansion and contraction is provided for by means of pipes with bends of large radius.

The boiler feed piping is arranged on a ring system so that, in case of an accident to any portion of the piping, the boilers may be supplied by the duplicate connection.

All of the valves used in both high and low pressure piping are made with adjustable wedge gates and with bronze seats.

The amount of coal required for the operation of the power station during the winter months is about 700 tons per day. This coal is brought to the power station dock in barges, from which the coal is unloaded by means of a ton-and-a-half clam-shell bucket operated by a hoisting engine. This bucket is elevated about fifty feet to the top of the coal tower, where the coal is discharged into crushers, which break it to a size suitable for use in the automatic stokers. The coal then drops into weighing hoppers, where it is weighed before going to the boilers. It is then elevated by means of conveyors to the top of the tower and is delivered into bucket conveyors which run the entire length of the boiler room, and these conveyors deliver to three separate coal bunkers built above the upper tier of boilers.

After the coal is burned, the ashes drop to the basement through rectangular cast iron pipes and, by opening valves at the lower ends of these pipes, the ashes are delivered into iron cars and an electric mining locomotive is used to pull these cars to a point at the east end of the building, where they are dumped into a hopper which loads the ashes into a line of bucket conveyors. These elevate and carry the ashes across the bridge to the tower containing ash storage bins. When a sufficient quatity of ashes has accumulated, these bins are emptied into barges which transport the ashes away.

A movable coal tower, provided with belt conveyors, has recently been installed in addition to the apparatus already described. The movable tower obviates the necessity of moving the barge from time to time as coal is unloaded, thereby saving much time and labor.

The engine room is provided with an electric crane which is capable of handling the heaviest single piece of machinery installed. The crane has one fifty-ton and one fifteen-ton hoist.

The eight Westinghouse alternators at the 74th Street power station were, at the time of their erection, the largest engine-driven dynamos that had ever been built.

The rated output of each alternator is 5,000 k.w., but they are designed to carry an overload of fifty per cent. for two hours with a temperature rise not to exceed 55° C., and in service they are frequently loaded to this amount for short spaces of time. The accompanying diagram shows a representative load at the 74th Street power station, taken on January 18, 1904, on a cold day when, in addition to the power for moving cars, a considerable quantity of power was required for car heaters. The diagram indicates that, during rush hours, the eight alternators were delivering to the line from 40,000 to 47,000 kilowatts, and that, in addition, from 5,000 to 6,000 kilowatts was being supplied by the Kingsbridge station of the Metropolitan Street Railway Company.

In order to meet the increasing demands for power caused by the rapidly increasing traffic, a 5,500-kilowatt alternator, driven by a Parson's steam turbine, is soon

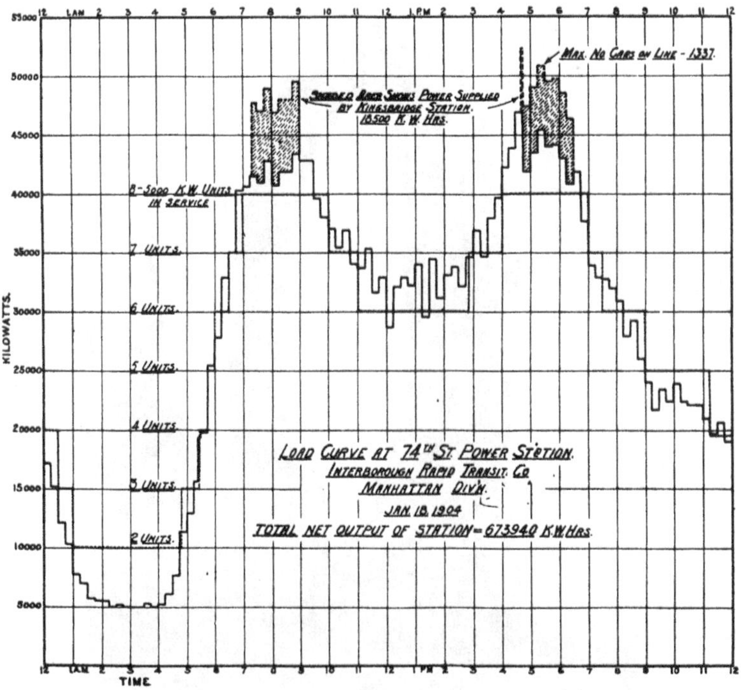

to be installed at the 74th Street station, and additional power will also be transmitted from the new Interborough power station at 58th Street, as soon as this station is in operation.

Reference to the load diagram shows that, between the hours of four and six-thirty in the morning, the load increased very rapidly from 5,000 to 40,000 kilowatts. Consideration of this quickly rising load was one of the determining factors in the choice of units of 5,000 kilowatts output. Even with units of this size, it is necessary to start one every twenty minutes in order to anticipate the morning load.

External Armature Frame of 5000 Kw. Alternators

The external armature frame of the alternator is forty-two feet high, and is made in six sections bolted together. This frame is bored to receive the laminated steel plates which form the armature core.

The winding is three-phase, and, at the normal output of 5,000 kilowatts at 11,000 volts with non-inductive load, the current per phase is 263 amperes. The normal speed is seventy-five revolutions a minute, and the current generated has 3,000 alternations per minute, or twenty-five cycles per second. The armature winding consists of insulated copper bars placed in partially closed slots, there being four slots per phase per pole and three bars in each slot. The insulation of the armature winding was subjected to a test of 25,000 volts alternating for thirty minutes before each machine was accepted by the purchaser.

The forty field poles are built up of laminated steel plates secured to the periphery of the steel plate fly wheel, and the outside diameter of the poles is thirty-two feet. The fly wheel effect of this rotating element is 370,000 pounds at 11.7 feet radius. The field windings consist of copper straps wound on edge, one layer deep, with insulating material cemented between the turns.

Copper wedges are driven into place between adjacent pole tips after the coils have been put on. These copper wedges serve to hold the coils in place and also act as a magnetic damper to check any tendency toward variation in angular velocity.

The exciting current required in the field coils when the armature is delivering full rated output, is 225 amperes at 200 volts. A field rheostat, with motor driven face plate, is provided with each alternator for regulating the potential. The insulation of the field coils is designed to withstand a test of 2,500 volts, alternating, for one minute.

The electrical efficiency of the alternator, determined by shop tests, is 96.68 per cent. at half load, 97.97 per cent. at full load, and 98.15 per cent. at 25 per cent. overload.

Four exciter generators are installed for supplying current to the alternator fields. Each generator is of 250 kilowatts output, or 1,000 amperes at 250 volts.

Each generator is direct connected to a 300-H. P.

tandem compound engine. The exhaust of these engines is piped to two motor-driven jet condensers.

The switching apparatus is placed on two galleries running longitudinally with the building and built against the partition wall between the engine and boiler rooms. The upper gallery is used for feeder switches and compartments containing the group bus bars, to which the substation feeders are connected. On the gallery below are built the alternator oil switches, and, in the double floor of this gallery, runways are constructed with brick partitions for the main bus bars. The instrument panels are also placed on this gallery at a point near the centre of the building, and the exciter and auxiliary switchboards are located at either side, so as to be near the operator on the same gallery. A third gallery, built below these two galleries, is used for the accommodation of the feeder cables. The General Electric Form "H" oil switch is used throughout, and, even under the most severe conditions in service, has opened 11,000 volt circuits without trouble of any kind.

The current from each alternator passes through a 500-ampere oil switch which is provided with an overload time-limit relay operating at the end of three seconds at about three times full load current. This oil switch is also provided with a reverse current time-limit relay which operates in three seconds at six-tenths full load current.

From this oil switch connections are made to two similar oil switches, without automatic relays, which are connected respectively to the two sets of main bus bars. These bus bars consist of stranded copper cable of 1,000,000 circular mils section, insulated with $\frac{9}{32}$ of an inch of rubber compound containing thirty per cent. fine Para. These cables are supported on porcelain insulators which are tested to 40,000 volts. Each cable is installed in a brick runway constructed under the floor of the switchboard gallery. A removable floor of two-inch slate is placed above these bus bars.

The feeder cables for each substation are provided with a group bus bar placed on the upper switch gallery.

For supplying current to each group bus bar, two 800-ampere oil switches, without relays, are installed, each being connected to one of the two main bus bars.

For each feeder cable, a 300-ampere oil switch is installed. This is provided with an overload time-limit relay set to operate at 300 amperes in about two seconds.

Space is provided on the gallery for six feeder oil switches for each substation, these switches being all connected to the corresponding group bus bar.

In all these oil switches each phase of the circuit is isolated in a brick compartment, each compartment containing two brass cylinders partially filled with oil. A copper rod runs through a stuffing box at the top of each cylinder and makes contact below the surface of the oil, so that, whenever a switch is opened, each phase is opened at two different points. An electric motor is employed to operate the switch, this motor being controlled by a miniature switch on the controlling board. The oil switch is very quick in its action, as the upward or downward movement is accomplished by means of compression springs, while the motor serves to follow up and compress these springs for the next movement of the switch.

A 110-volt storage battery supplies current for the operation of the oil switch motors and the circuits to these motors are so arranged that the current is automatically cut off the motors at the end of each operation.

Two indicating lamps, placed on the operating board, are provided for each oil switch and are wired to contacts on the switch so that a red light indicates that the switch is closed, while a green light indicates the switch open.

The operating switches for controlling the alternator oil switches are placed on a controlling board and are arranged with miniature bus bars which indicate to the operator, diagramatically, the connections of the circuits. A similar controlling board is provided for operating the feeder oil switches.

On the instrument panels, each alternator is provided with three horizontal, edgewise ammeters, one voltmeter, one indicating wattmeter, one power factor indicator, and one recording wattmeter.

A synchronism indicator is placed on the instrument board and connections are made to this indicator by means of synchronizing plugs and receptacles. In synchronizing an alternator with others already in operation, the indicator pointer moves either to the right or to the left, indicating thereby whether the alternator is running too fast or too slow. When the proper speed has been reached, the pointer is stationary. The oil switch is closed just before the pointer approaches the central position, for, at this time, the speed of the alternator which is about to go into service is approximately the same as the speed of the other machines and the alternations are also in synchronism.

The use of this indicator greatly simplifies the operation of synchronizing and is found to be preferable to the use of synchronizing lamps.

Each feeder cable is provided with three ammeters, one for each phase, placed on the instrument board of the power station.

For supplying 500-volt, direct current to the lights and auxiliary motors in the power station, three 800 kilowatt, six phase, rotary converters are installed. The coal and ash handling machinery, boiler feed pumps, exciter condensers, and crane are operated by current from these converters.

The engine room is lighted by 1,500 sixteen candle-power incandescent lamps fastened to the columns and to the under side of the roof trusses. The lamps are of the 130-volt, railway type, connected four in series.

For the general illumination of the boiler room, arc lamps, supplied with 500-volt direct current, are provided, supplemented by incandescent lamps around the gauge glasses and boiler feed pumps.

DATA OF POWER STATION OPERATION

The men employed in the operation of the power station may be classified as follows:

Superintendence and office force		14
Chemist		1
Men employed on electrical apparatus:		
Switchboard attendants	9	
Dynamo tenders and cleaners	18	
Electrical repairmen	7	
	—	34
Men employed on engines:		
Steam engineers and assistant engineers	17	
Oilers, wipers and oil system attendants	53	
Machinists and helpers	20	
Steam fitters and helpers	7	
Blacksmith and helper	2	
	—	99
Men employed in boiler room:		
Boiler room engineers and attendants	20	
Boilermaker, helper and cleaners	19	
Stoker operators and assistants	57	
Coal handling machinery attendants	13	
Ash handling machinery attendants	10	
Pump men	9	
Mason and helper	2	
	—	130
Laborers and foreman		16
Janitors and doormen		6
Total		300

In the operation of the power station the proportion of expense is divided as follows:

Operating expenses	92.25	per cent.
Maintenance expenses	7.75	"

The operating expenses are divided as follows:

Coal, at $3.15 per gross ton	71	per cent.
Labor and miscellaneous expenses	20.88	"
Water	7.7	"
Oil, waste, rags and grease	.42	"

The following figures are given to show a comparison between the cost of operation during the summer and winter months. Comparing the watt hours per ton-mile, it will be seen that during the winter months the power required is about twenty-two per cent. greater than that required during the summer months, the difference being due almost entirely to the use of electric heaters on the cars:

	Summer Months, Average for July, Aug., Sept., 1903	Winter Months, Average for Jan., Feb., March, 1904
Average K.W. hours (net) per day delivered......	431,197	595,996
Average lbs. coal per K.W. hour (net)	2.649	2.610
Average lbs. water per K.W. hour (net)......	23.21	24.17
Average lbs. water per lb. coal	9.00	9.27
Average K.W. hours (net) per car mile	2.566	3.35
Average Watt hours (net) per ton mile...........	100.13	122.68
Average K.W. per car from maximum fifteen-minute readings	28.99	39.04

HIGH TENSION CABLES

The power is transmitted from the 74th Street power station to the seven substations through three-phase cables placed in underground conduits. The longest distance which current is transmitted is about 5.3 miles, and the total number of miles of cable installed is 130.2.

Each cable has three 000 B&S gauge stranded conductors, each insulated with $\frac{7}{32}$ of an inch of paper saturated with insulating compound. The three conductors are laid together with jute and surrounded by an envelope of $\frac{7}{32}$ of an inch of paper insulation with a lead sheath $\frac{9}{64}$ of an inch thick over all. This cable is 2.64 inches outside diameter and weighs 8.8 pounds per foot.

An insulation resistance of not less than fifty megohms per mile, at sixty degrees Fahrenheit, is required and, before leaving the factory, each cable length is subjected to a potential of 30,000 volts, alternating, for thirty minutes between conductors and between each conductor and the lead sheath. After the cables are laid and jointed, they are again subjected to high potential tests consisting of 30,000 volts, alternating, for thirty minutes between conductors and 22,000 volts, alternating, between each conductor and the lead sheath.

The cable conduits are built of single vitrified clay ducts of five inches square outside and eighteen inches long, with a hole three and one-half inches in diameter. The ducts are laid on a bed of concrete four inches thick, with three inches of concrete at the sides of the group and four inches over the top. As an additional protection, two-inch creosoted planks are laid above the top layer of concrete. The layers of ducts are flushed with thin cement as they are laid. Cable manholes, six feet by eight feet inside, are built from 200 to 500 feet apart, the length depending upon local conditions.

The joints in the cables are made in the cable manhole and each cable is wrapped with asbestos cloth one-fourth of an inch in thickness, secured in place by bands of galvanized iron. This wrapping is for the purpose of localizing any trouble which may occur, so that a burn-out in one cable will not be communicated to adjacent cables.

Where the cables cross beneath the Harlem River, submarine cable is installed, each conductor of which is insulated with $\frac{9}{32}$ of an inch of forty per cent. pure Para, the three conductors being laid up in jute and surrounded by a lead sheath $\frac{8}{32}$ of an inch in thickness, and the whole being protected by two steel tapes wound in reverse, which, in turn, are covered by tarred jute.

SUBSTATIONS

The seven substations are built on a uniform plan, the buildings being in all cases approximately fifty feet wide and one hundred feet long.

The rotary converters are placed on the first floor, which is level with the street. The basement is taken up by converter foundations, with space for blowers.

A gallery seventeen feet six inches above the first

Interior of Substation Manhattan Division

floor extends around the four sides of the building. The switchboard is located across the front of the building on this floor, and the step-down transformers are placed along the side-galleries.

There are two floors above the main floor, which were built for the possible installation of storage batteries.

The buildings are all of steel construction and are fireproof throughout. The first story of the front wall is built of granite, above which the wall is of fire-flashed brick.

In each substation, with the exception of Substation No. 5, there are six 1,500 kilowatt Westinghouse rotary converters. Substation No. 5 has five converters, which makes a total of forty-one converters in all. These converters have twelve poles and are arranged for operation at 250 revolutions a minute on a three-phase circuit having twenty-five cycles per second. Each converter has a normal output of 1,500 kilowatts, namely, 2,400 amperes at 625 volts, the alternating current being 2,350 amperes per terminal at approximately 390 volts. The guaranteed efficiency of the converters is 93.5 per cent. at half load, 95.75 per cent. at full load, and 96 per cent. at 25 per cent. overload.

The fields of the rotary converters are provided with both series and shunt windings, but thus far the shunt windings only have been in use, as the drop in voltage on the heaviest loads is so slight as not to require the use of the series windings.

For the purpose of starting the rotary converters, a motor-generator set is provided in each substation. The converter is started with direct current from this starting set and brought to synchronism by means of a synchronism indicator. The oil switch is then closed and the converter fields are adjusted so that it takes its share of the load.

Three 550-kilowatt transformers are installed with each rotary converter to reduce the voltage from 11,000 to 390 volts. These transformers are of the air-cooled type. They are placed on the gallery above the converters and over an air duct made of cement on expanded metal.

Two motor-driven, direct-connected blowers in each substation furnish air at about one ounce pressure per square inch under the transformers.

The guaranteed efficiency of these transformers is 97 per cent. at half load, 97.75 at full load, and 97.70 at 25 per cent. overload.

The transformers are tested with 25,000 volts, alternating, for thirty minutes.

The 11,000-volt, three-conductor cables which supply

current from the power station, enter the substations through the basement wall and each cable terminates in a copper end bell. From the end bell, three single conductor cables run to an oil switch which has instantaneous overload and reverse current relays. These switches are connected to a set of three bus bars provided with knife switches between the connections of each set of incoming feeder cables, so that the cables may be operated separately or in multiple.

An oil switch for each rotary converter is connected to the bus bars, and single conductor cables run from these oil switches to the 550-kilowatt transformers. The oil switches are placed on a gallery in the front of the substation, directly above the direct-current switchboard.

The direct-current switchboard carries only the positive bus bar and its connections. There is one panel for each rotary converter, which has mounted upon it the circuit breaker, ammeter, double-throw switch, recording wattmeter and converter starting switch. A spare converter panel is provided, so arranged that, in case of trouble on the circuit breaker, the current may be switched to the spare panel while the necessary repairs are being made.

Each direct-current feeder is provided with a panel having mounted upon it a circuit breaker, ammeter, double-throw switch, and a spare panel is provided so that repairs may be made on the circuit breaker when required.

The controlling switches for the operation of the 11,000-volt oil switches are arranged on a controlling board with indicating lamps and with miniature bus bars which indicate the circuits diagramatically.

The oil switch on each incoming feeder is provided with a relay which opens the switch automatically on reversal of current.

Each rotary converter is furnished with one 200-ampere ammeter, one 3,000-kilowatt indicating wattmeter, one power factor indicator, one recording wattmeter, and one time-limit relay for automatically tripping the oil switch on overload. These are all alternating current instruments. The time-limit relay is set at 240 amperes, or about 2.7 times full load current, and operates at the end of one and one-half seconds.

Each substation is provided with a twenty-five-ton

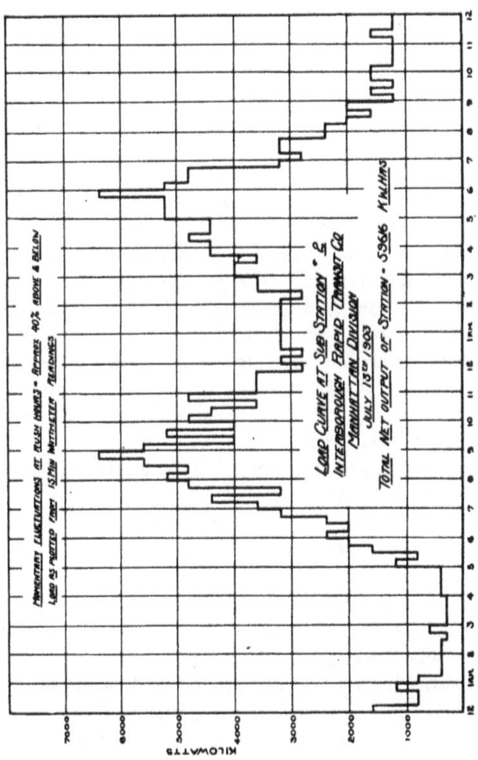

electric crane placed above the rotary converters, and two five-ton hand cranes, one placed above each transformer gallery

A representative twenty-four hour load on a substation is shown on the accompanying diagram. This load was obtained from fifteen-minute readings of the recording alternating current wattmeters. The momentary loads are about forty per cent. above and below these readings, due to the simultaneous starting of the six-car trains. In service the converters have succssfully carried as much as one hundred per cent. overload each, for short periods. The substation load in the winter is about twenty per cent. heavier than that shown, due to the current required for car heaters.

CONTACT RAIL AND FEEDER SYSTEM

The contact rail used on the Manhattan Division is a standard section of the Lackawanna Iron & Steel Company, weighing one hundred pounds per yard, rolled in sixty-foot lengths. It was considered advisable to use a heavy rail with low resistance in order to save the expense of additional copper cables, and the rail was made of a special composition in which the percentage of carbon and manganese was as low as possible. The resistance is about eight times that of an equal section of copper, while the resistance of a high carbon rail is about twelve times.

The average composition of the steel in this rail is as follows:

Carbon	.03	per cent.
Manganese	.341	"
Sulphur	.073	"
Phosphorus	.069	"

The rail is very soft and would be quite unfit for ordinary wheel wear.

The direct current feeder system is divided into sections, each section covering the distance between two adjacent substations. The contact rails for up and down-town tracks are connected with cross bonds and usually no additional copper feeder cables are required.

Current is supplied at each substation from the positive bus bar through a 10,000-ampere circuit breaker and switch. The division between the sections occurs opposite each substation and, at these points, a gap forty feet long is left between adjacent ends of contact rails so that the car wiring on a single motor car cannot span across from one section to the next. This gap is made by placing two strips of wood together with sections of Tee iron between and with the ends of the Tee iron one inch apart. The top of the Tee iron is on a level with the top of the contact rail.

At a point half way between substations, a similar break is made in the contact rail, and this break is bridged with a 6,000-ampere circuit breaker and switch which are normally kept closed. This circuit breaker at the centre of the feeder section enables the section to be divided into equal parts in case it is necessary to localize trouble.

This arrangement of the direct current feeder system utilizes to the fullest extent the conductivity of the contact rails and enables the load to be equalized between adjacent substations. Additional copper cables are provided whenever the drop in the contact rail, at a point farthest from the substation, exceeds a predetermined amount, based on the laws of economic operation. These additional feeders are made up of 1,500,000 C.M. bare copper cable, installed on glass insulators resting on the ties, the whole being covered with a wooden box made of two-inch plank.

The contact rail is installed on vitrified clay insulators about six inches square and four and one-half inches high. These clay blocks are fastened with Portland cement to malleable iron pedestals which are secured to the ties. The design provides a drip edge on the under surface of the insulating block. Malleable iron clips are clamped to the top of the insulating block, which serve to hold the contact rail in place with a generous allowance for both horizontal and vertical clearance.

The expansion of the contact rail is provided for by dividing it into 300-foot sections and providing three-inch expansion joints at the ends of these sections. At the center of the sections, special clips are used which anchor the rail firmly to the ties.

The rail joints are bonded with four solid bonds of

drop forged copper. The total sectional area of the bonds is such that the conductivity is slightly greater than that of the contact rail. Two of these bonds are riveted to the base and two to the web of the rail. The rails are held together by means of two short malleable iron splice bars.

The expansion joints are bonded with four extra flexible cables with copper terminals cast and drop-forged to the ends of the cables. The malleable iron splice bars are slotted and allow for a maximum movement of three inches.

At crossovers, turnouts and other points where the continuity of the rail is necessarily interrupted, cast iron inclines are secured to the ends of the contact rail to guide the contact shoes up to and down from the level of the rail.

A wooden guard strip is bolted to the outside of the contact rail in order to protect employees and also to guard against short-circuits.

For the negative, or return side of the direct current feeder system, the elevated structure and track rails are bonded together, thus forming a return path of great conductivity, so that the drop in the return circuit is very small. The track rails are ninety-pound high carbon rails bonded together with one 0000 bond per joint. At intervals of ninety feet, the track rail is bonded to the top chord of the longitudinal girder by means of a 000 flexible copper bond. The upper and lower chords of the longitudinal girders of the structure are bonded together with from one to four 0000 copper bonds, the number of bonds increasing toward the substations.

The structure and track return is connected, by means of bare 1,500,000 circular mils cable, to the negative bus bars in the substations.

CAR EQUIPMENT

In deciding upon the electrical equipment of cars to replace steam locomotive service on the Manhattan Railway, it was important to take advantage of the possibility of operating trains at increased speed, but, as increased speed demanded additional weight on the driving wheels, it was necessary to distribute the weight so as not to exceed the safe bearing strength of the

Manhattan Elevated Railway Train

structure. The "multiple unit system" of train control was adopted as the only system which fulfilled all the requirements, and two years' experience has demonstrated the wisdom of the choice.

The local train service on the Manhattan Division requires frequent stops, as the stations are about 1,775 feet apart. The time occupied by the stop at each station during the hours of heavy traffic averages about fourteen seconds. It was determined that the most economical speed for the Manhattan service was 13.5

miles an hour during the hours of maximum traffic. During the hours of lighter traffic, the time occupied by stops at stations is enough less so that trains are run at a schedule speed of 15 miles an hour.

Six-car express trains are run in the morning and evening on the center track of the Ninth and Third Avenue lines. The equipment for express service is identical with that for local service.

On account of the increased rate of acceleration and braking, extensive changes were required in the old cars and trucks, and some of the old cars were considered unsuitable for electric service and were retired from use. Four hundred and ninety of the best of the old cars were altered for service as motor cars. In these cars a motorman's cab was built in each end, and the platforms were rebuilt, new bolsters put on, and the car sills re-enforced. Three hundred and sixty new motor cars were purchased, so that there are at present 850 motor cars of the Manhattan type in service. Four hundred and twenty-six of the old cars were altered for trailer car service by re-enforcing the sills, rebuilding the platforms and supplying new bolsters. The vacuum brakes were removed from all of the old cars and replaced by Westinghouse automatic air brakes, the old link and pin couplers were replaced by Van Dorn draw bars with automatic couplers, the Pintsch gas system was removed and incandescent electric lamps substituted, and the steam heaters were replaced by electric heaters.

Thirty-six open cars with cross seats for eighty passengers have been purchased and have proved very attractive to the public for summer use.

Owing to the great increase in traffic, 250 of the new cars which were built for the Subway service have been put into service on Second Avenue.

The new motor car bodies were built to conform in general design and appearance to the best type of cars developed in the steam service of the road. They are, however, made stronger in construction and are about 1,400 pounds heavier. Each car provides seats for forty-eight passengers, with eight cross seats at the centre and the remaining seats along the sides of the car. Sliding doors are provided at the ends of the cars. An interesting feature of these cars is the arrangement of the motormen's cabs, which are built inside the car with folding doors and with hinged seats so that, when not in use by the motorman, the cab doors are folded against the ends of the cars, thus enclosing the controllers and air brake valves and the seats lowered and occupied by passengers. Underneath the hinged seat, a motorman's seat has been arranged, supported on a sliding bracket so that, when the cab is to be occupied by the motorman, the hinged seat is first raised to a vertical position against the rear wall of the cab and then the bracket seat dropped into position for the motorman's use.

The car platforms are provided with swinging gates operated by a patent gate mechanism. Folding safety gates are also provided between the platforms of adjacent cars.

The interior finish of the car is of mahogany, and the ceilings are of three-ply whitewood or birch, finished in a light cream color with stenciled gold leaf ornaments. There are twelve windows along each side of the car, spaced so that the posts between windows come opposite the divisions between side seats. Pantosote curtains, with rolling fixtures, are provided in preference to the wooden shutters formerly used.

A headlight, consisting of an incandescent lamp in a parabolic reflector, is placed at each end of the motor car at the centre of the car roof. At each side of the headlight is placed a designation signal lantern or "marker," illuminated by an incandescent lamp. The four sides of the marker are painted in white, red, green, and orange, with glass bull's eyes of like color. The colors which are visible from the forward end of the car indicate the destination of the train. The marker is turned to its proper position by means of a spindle extending through the car roof.

The electric heaters are arranged in three separate circuits, which require eight amperes each. The heater circuits are controlled by three quick-break, knife switches, with enclosed fuses mounted on a slate panel in an iron box placed at one end of the car. The heaters are placed beneath the side and cross seats, so as to give an even distribution of heat throughout the car, and great care has been taken in the wiring to ensure against the possibility of fire originating from loose connections or damaged insulation.

Each motor car is provided with twenty-five 10-candle-power incandescent lamps for interior illumination. The lamps are placed just above the space for advertising signs along the sides of the car, and five lamps are placed at equal distances along the centre of the dome ceiling. The switch and enclosed fuses are mounted on a slate panel in an iron box of the same design as that for car heater circuits. The wiring is insulated for 2,000 volts and runs in wood moulding.

The motor truck was designed by the engineers of the Manhattan Division, following closely the master car builder's type of construction. The trucks have a six-foot wheel base. The truck frame is made of 4" by 3½" by ⅝" rolled angle iron weighing 14.6 lbs. per foot. This section was used in preference to a forged rectangular bar, as it not only reduces the weight for equivalent strength, but also minimizes the expense of construction.

The contact shoes which transmit current from the third rail are supported at each side of the truck by an oak beam, the ends of which rest on the coiled spring seat castings. The contact shoes are of cast iron and weigh about thirteen pounds each. They are hung on a pair of cast steel links which allow free vertical movement. A flexible copper cable is connected to the contact shoe for carrying the current. The casting which supports the links from the oak beam is arranged for vertical adjustment of two inches to allow for wear of wheels.

The wheels are thirty-four and one-fourth inches in diameter, and allowance is made for a reduction of two inches in tire thickness. The rolled steel tires are secured to a cast iron centre by means of two steel rings of the Mansell type bolted in place with the ends of the bolts riveted over. The wheel has eight radial spokes, oval in section, which form a part of the wheel centre casting.

An interesting feature of these wheels is the method of attaching the driving gear to the extended hub of one wheel by means of a shrink fit. The extended hub is turned to a diameter $\frac{1}{100}$th of an inch larger than the bore of the gear. The cast steel gear, which is in one piece, is then heated until the expansion of the metal allows the gear to slip on over the hub easily. This method of attachment greatly reduces the liability of broken axles, for, as one wheel is driven directly from the gear, the tortional strain in the axle is reduced one-half. The elimination of the key seat removes an element of danger, for breakage of axles usually starts at this point of weakened section. The solid gear does away with the necessity of gear bolts, which are always liable to become loose.

When it becomes necessary to remove a gear, it is expanded by heat until it slips off easily from the extended hub, then the wheel at the other end of the axle is expanded in a similar manner and pressed off with light pressure, so that the metal is not scored and the wheel can be replaced after a new gear is put in. After a service of two years, not one of the gears has become loose and not one axle has broken.

SLEET AND SNOW SCRAPERS

After experimenting with many devices, which promised success and realized failure, the Manhattan management adopted a cast steel sleet scraper, which, in connection with the installation of a bus line cable, has enabled the trains to run without delay through sleet and snow, as demonstrated by last winter's service.

Each motor car is provided with four scrapers, which are fastened to the oak beam supporting the contact shoes. The scraper consists of six parallel blades set diagonally across the rail at an angle of 60°. Each blade is ⅜ of an inch thick at the bottom and increases to $\frac{9}{16}$ of an inch at the top. The steel casting, of which the blades form a part, is supported by guides which allow a vertical movement of 2½ inches. A flat steel spring is arranged to press downward on the scraper casting with 140 pounds pressure, which, added to the weight of the casting, makes a total pressure of 165 pounds on the third rail. A cam with a wooden lever attached is provided for raising the scraper by hand. Tripping blocks are placed along the road, which, when raised into position, strike the cam levers and trip the scrapers down as the train passes. It is necessary to

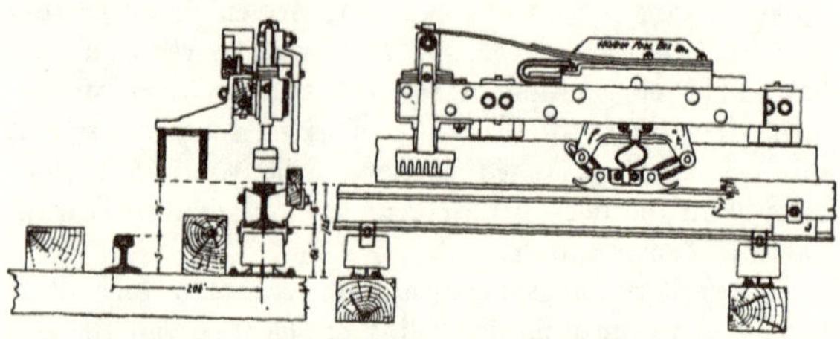

Contact Shoe and Sleet Scraper, Manhattan Division

have the third rail in good alignment with all expansion joints beveled in order to avoid breaking the scrapers.

The bus line serves to connect all of the contact shoes together in multiple, so that, in case of snow or sleet, even if the forward shoes do not make contact, the rear shoes will supply current to all motors throughout the train through the bus cable. The bus line also prevents the "flashing over" of the motors that frequently results without its use, when a contact shoe jumps and breaks the circuit while the car is running in multiple at high speed.

In order to protect the bus line from excessive currents, two 750-ampere copper ribbon fuses are placed in the circuit under each motor car.

MOTORS

Each six-car train is supplied with eight motors placed two on a truck on each of four motor cars. There are used in the equipment 1,674 motors, known as General Electric Type No. 66. Each motor is rated at 125 H.P., based on a temperature rise of 75° C. above the surrounding air, after a one-hour run at full load, the air temperature not exceeding 25° C.

The motor shell or frame is made of cast steel, in one piece. The ends of this shell are bored large enough to permit the removal of the armature, after taking out the frame heads which carry the armature shaft bearings. The bearings are lubricated by the use of oil and waste in a manner similar to that of a car axle journal. The four field coils are wound with strip copper insulated with mica and asbestos. These coils are encased in metal spools, which protect the coils from moisture and from mechanical injury, and facilitate the radiation of heat. The pole pieces are made of laminated iron, bolted to finished surfaces on the motor shell. The armature conductors are separately insulated with mica and assembled in sets of five, which are protected with mica wound on with an outside layer of tape.

The insulation is subjected to the following tests before leaving the factory:

	Volts, A. C.
Armature windings, between conductors and core.	3,500
Commutator, between segments and shell........	4,500
Commutator, between adjacent segments	500
Field coils, between windings and frame..........	500

The gears are made of cast steel in one piece, bored to shrink on the extended hub of the wheel. The pinion is made of hammered steel with tapered fit on the armature shaft, and the teeth are five inches wide.

The weight of the motor complete, with gear and gear case, is 4,422 pounds. The gear alone weighs 313 pounds and the gear case 144 pounds.

The General Electric Type "M" control apparatus, in use on the motor cars, embraces a number of electrically operated switches, called "contactors," and an electrically operated reverse switch called the "reverser." The contactors and reversers are operated simultaneously on all the motor cars of the train from the master controller, which is turned on and off by the motorman. A nine-wire cable, with suitable couplers between the cars, runs throughout the train, connecting the master controller and the contactors and reversers.

The successive operation of the contactors cuts out the resistance boxes and makes the series and parallel combination in a manner identical with the ordinary hand controller in street car use, and the position of the master controller indicates to the motorman the position of the contactors on all cars.

The rate of acceleration of the train is, therefore, in the immediate control of the motorman. The master controller handle is so arranged that, if the motorman removes his hand, the current is at once cut off from all motors in the train.

The reversers are operated by the reverse switch of the master controller, and are interlocked so that they can only be thrown when current is off the motors.

A cut-out switch is provided on each motor car, so that the contactors and reverser on this car can be cut out without affecting the operation of the apparatus on the remaining cars. The operating coils of the contactors are wound for working at a maximum potential of 600 volts without undue heating, and will successfully operate at 300 volts. At 550 volts line potential, the current required for the operation of the contactors on each car is about 2.5 amperes.

The weight of the complete control apparatus on each motor car is 1,733 pounds, exclusive of the wiring.

In service, this control apparatus has proved to be very reliable and the cost of maintenance is much less than was anticipated. As a rule, the only parts that require renewal at the end of one year's service are the removable copper tips that make and break the motor current on the contactors. The contactors are hung in a row on the under side of the car body, where they are easily accessible for inspection, which is made once every three days.

The greatest care has been taken in the installation of the car wiring to reduce the fire risk to a minimum. All the wiring is done with stranded cable and the connections between the trucks and car body are extra flexible. The main motor cables are insulated with $\frac{5}{64}$ of an inch of rubber having not less than thirty per cent. Para and covered with double weather-proof braid.

The entire under surface of the car is lined with one-fourth inch asbestos board, and the cables are all placed below this fireproof sheath in continuous moulded forms of the same material.

The main wiring is protected against undue heating, from accidental short-circuit, by 400 ampere enclosed fuses placed in the circuit immediately above each contact shoe. These fuses are protected from the weather by oak boxes lined with asbestos and with ventilating spaces. The fuses have been developed after tests covering a year or more and are able to successfully open a dead short-circuit of 15,000 to 18,000 amperes, with almost no noise and little smoke.

A copper ribbon, 400-ampere fuse, placed in an asbestos-lined box under the car, serves as an additional protection to the motors. In case of overload, this fuse will blow before the contact shoe fuses, as there are at least two of them in multiple for each motor car.

During construction, all the car wiring was subjected to a test of 2,000 volts alternating for one minute, and the same test is applied to the wiring each time the car is in the shop for general repairs.

The cars are equipped with Westinghouse automatic air brakes. The air is supplied by motor-driven compressors. One compressor is placed under each motor car. The compressors are simultaneously stopped and started by automatic governors, so that the air supply in the storage reservoirs is kept between the limits of 85 and 95 pounds pressure.

A balance wire is run throughout the train, so that,

whenever one governor acts, all the compressor motors are started, and they continue to run until the last governor has cut out. The storage reservoirs on all cars in the train are connected by a reservoir pipe in order to equalize the pressure. The weight of each compressor is about eight hundred pounds.

PRIVATE TELEPHONE SYSTEM

In order to facilitate communication, a private telephone system has been installed, comprising about four hundred telephones for the Manhattan Division, and two hundred additional telephones will soon be installed for the Subway Division. A telephone is provided in the ticket office of each passenger station, and all executive offices, switch towers, car shops, substations and power-house are also provided with telephones.

These telephones are connected on what is known as a "common battery" system, with a central exchange located near the centre of the system at 99th Street and Third Avenue.

It is of interest to know that the length of line from the exchange to the most distant instrument is about nine miles, which is considerably more than is usual on common battery systems where the telephone wires are in lead covered cables.

The telephones in passenger station ticket offices are bridged together, two on one circuit, but all other telephones have a separate wire to the exchange board.

The telephone wires are No. 19 B&S gauge, with dry paper insulation made up into lead covered cables. The cables contain from five to one hundred pairs of wires each, and are suspended from the elevated structure on span wires supported at the structure columns. It is necessary to use every precaution to guard against the vibration or bending of the lead sheath, as it is found that continued movement of any kind soon causes crystallization of the lead.

For the Subway Division, the telephone cables are run in underground ducts in the walls of the subway.

The telephone exchange is provided with positions for six operators, and the present service requires four operators during the busiest hours. The telephone exchange is complete and modern in every particular, and the same conveniences for the telephone operators have been provided as are found in the large exchanges throughout the city.

This telephone system is found to be of great service in the transaction of Company business, and it is especially valuable in that it provides a means of prompt communication to all points on the line whenever accidents or delays to traffic of any kind occur.

LIGHTING SYSTEM

A complete system of electric lighting has been installed at the passenger stations, car barns, and other buildings belonging to the Company, which takes the place of the gas service formerly in use.

The lighting system is supplied by 2,500-volt, twenty-five-cycle, alternating current, distributed through three-conductor cables installed on glass insulators on the structure. The current is supplied by transformers which reduce the voltage from 11,000 to 2,500 volts. A set of three of these transformers of 79 kilowatts each, is installed at each of the seven substations to supply current for lights at the passenger stations and shops located nearest to this substation.

At each passenger station, a transformer is located which lowers the voltage from 2,500 to 120 volts. The neutral point of the secondary winding is permanently grounded to avoid possible danger from shocks due to accidental grounding of the primary winding. Each transformer is provided with a 2,500-volt switch and fuses. The loads on the three phases of the feeder cables are balanced by connecting the transformers alternately across different phases.

Each passenger station is provided with about 100 sixteen candle-power, sixty-four watt, lamps. These lamps are wired on circuits of approximately ten lights each, these circuits being controlled from a panel switchboard placed in the waiting room. All of the wiring is enclosed in loricated conduit pipes.

For the illumination of the yards, clusters of ten sixteen candle-power incandescent lamps are used. These clusters are placed with a twenty-seven-inch reflector at the top of a three-inch iron pipe pole twenty-five feet above the tracks. These lights are wired five in series and fed from the third rail.

The lighting system comprises about 25,000 incandescent lamps in all, of which 14,000 are at the passenger stations, the remainder being used in shops, inspection sheds and other buildings, and in the illumination of the yards.

Electrical Equipment of Interborough Rapid Transit Company: Subway Division

ON February 21, 1900, a contract was executed with John B. McDonald for the construction of the Subway, which is now practically completed. This contract was carried out by private enterprise, with financial backing guaranteed by the city, and a provision was made for reversion of the property to the city at the end of fifty years free and clear of all encumbrances.

The formal opening of construction work was made at the City Hall Park on March 24, 1900. The contract covered the construction of a four-track subway from City Hall to 96th Street, a distance of 6.71 miles; an extension on the west side of the city to Kingsbridge, made up of 3.85 miles of three-track construction, and three miles more of two-track construction; and an extension on the east side of the city to Bronx Park of 2.89 miles of three-track and 4.24 miles of two-track construction, making a total of 20.69 miles of subway and elevated structure, providing for 63 miles of single track. In this distance there are 10.6 miles of subway built with steel columns supporting the street, 4.5 miles of concrete arch construction, and 5.0 miles of elevated structure—the latter construction being employed only at the outlying ends of the line and where the contour of the ground made the elevation of the tracks necessary in order to obviate steep grades. The line was divided into nineteen sections, and the construction of these sections was sublet to various contractors.

The contract also covered the furnishing and instal-

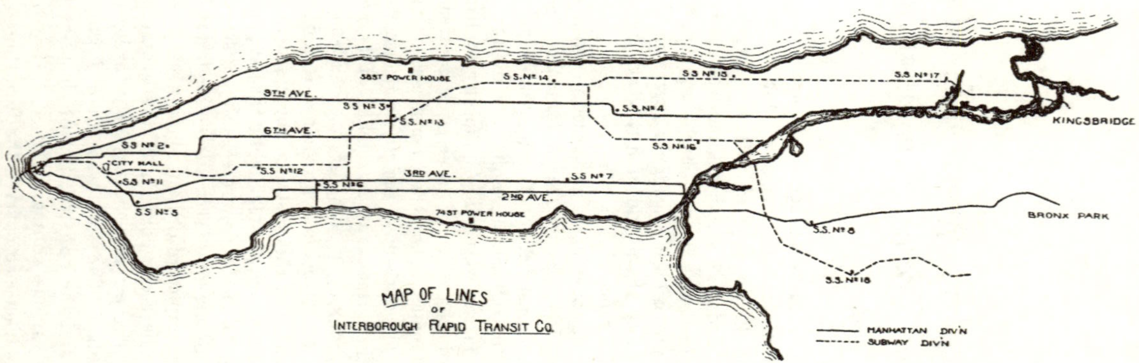

MAP OF LINES OF INTERBOROUGH RAPID TRANSIT CO.

lation of power station, substations, power transmission lines, and rolling stock for the operation of the line.

The amount of power for the operation of the trains is based on the number and weight of the cars to be operated and the character of the service to be rendered.

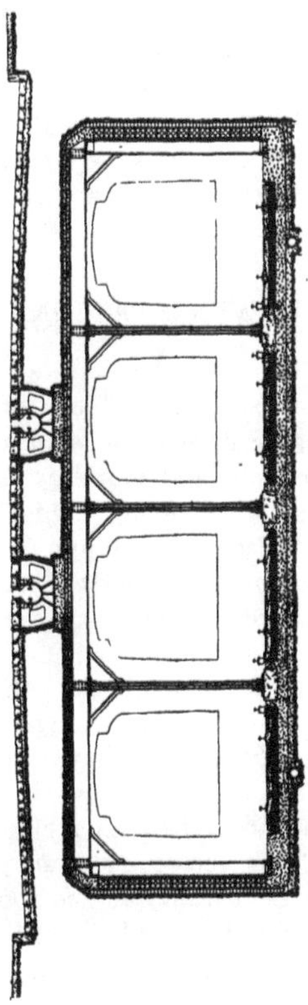

Section of Four-track Subway

The local service will consist of five-car trains, which will run on one-minute headway on the main part of the line, with alternate trains to each branch. The contract calls for a speed of fourteen miles an hour from end to end of the line, but it is hoped that this speed will be exceeded by two miles an hour. The average distance between stations is 2,300 feet.

The express service will consist of eight car trains, which will run from City Hall to 96th Street with stops at 14th, 42d and 72d Streets. These trains will run on a two-minute headway to 96th Street, alternate trains going to the east and west side branches. The average distance between express stations is 8,500 feet.

The combined express and local service requires about seven hundred cars, and the power required when the maximum number of trains are on the line is estimated to be 52,300 kilowatts.

POWER STATION

The power station is located on the North River, between 58th and 59th Streets. It is built to contain eleven 5,000 kilowatt engine driven alternators, which furnish power for Subway cars, and four 1,250 kilowatt turbine driven alternators for supplying lights in the Subway. The building is 201 feet in width and 690 feet in length. The property extends 110 feet west of the present building to the North River, leaving space for two additional units.

The station differs from all the other large power stations in New York in the arrangement of boilers, which are all on one floor placed in two rows, with centre firing aisle.

Six brick stacks are to be built for the present building in line with this centre aisle, supported on steel columns, which extend nearly to the roof of the building. The stacks, which are of Custodis hollow brick, are fifteen feet inside diameter and 225 feet above the grates.

The foundations for building and machinery rest on solid rock.

The exterior walls are finished in pressed brick of grey color, with terra cotta trimmings and with green roof tile.

A face work of cut granite extends around the power house up to the water table, above which face brick of a light grey color is used. Wire glass is used in the windows which have cast steel frames. All flashing and window capping is made of copper.

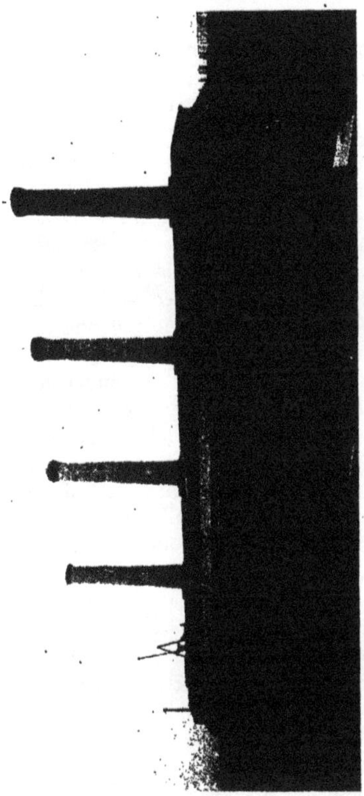

58th Street Power Station, Subway Division

The power station machinery is arranged in six units, each unit consisting of one stack, twelve boilers and two engines.

The nine Allis-Chalmers engines ordered are of the

same type as those in the Manhattan station. The two horizontal high pressure cylinders are forty-two inches in diameter, and the two vertical low pressure cylinders are eighty-six inches in diameter.

The high pressure cylinders are supplied with poppet valves instead of Corliss valves, so that superheated steam may be used. Each engine is rated at 7,500 indicated horse-power when operating at best efficiency at seventy-five revolutions a minute, with 175 pounds steam pressure at the throttle and twenty-six inches vacuum.

The guaranteed steam consumption is twelve and one-fourth pounds of dry steam per indicated horse-power per hour, at 7,500 indicated horse-power at seventy-five revolutions a minute, twenty-six inches vacuum and 175 pounds steam pressure, this steam consumption to include all steam used in jackets and reheaters.

Each engine is provided with an independent condensing equipment, consisting of one compound vertical crank-and-flywheel circulating pump, one crank-and-flywheel dry air pump, and two condensing cones. These cones are placed one immediately below each of the low pressure cylinders, at a height of about forty feet above extreme high water in the discharge tunnel.

The pumps, together with the boiler feed pumps, all of which are steam-driven, are placed in a continuous row adjacent to the division wall between engine and boiler rooms.

The condensing water conduits are on the 58th Street side of the power station, and the intake tunnel is provided with movable screens. Movable screens are also provided for the suction pipes of the circulating pumps.

The hot well seals for the barometric condensers are on the basement floor under the engine room, from which the water is taken through tile piping to the overflow conduit in 58th Street.

The boiler plant for the present installation consists of sixty Babcock & Wilcox horizontal water tube boilers, each having 6,000 square feet of heating surface. These boilers are all placed on one floor in batteries of two boilers each, and are supported on the steel columns of the building. Provision is made for superheaters.

Grates for hand firing are installed for the first thirty-six boilers and automatic stokers for twelve boilers. Each boiler has one hundred square feet of grate area.

A permanent gallery is built around the firing aisle along the front of the boilers, about ten feet from the floor, for water tenders and for convenience in cleaning the boilers. A hand crane is also provided above the firing aisle in order to facilitate construction and repairs.

Coal is deposited on the floor in front of the boilers through iron chutes from the coal bunkers above the boilers.

Provision is made for twenty fuel economizers for heating the feed water. These are to be placed immediately above the boilers with a main flue on the same level. The fuel gases from sets of three boilers pass through each economizer, with dampers and by-pass flues, so that the gases may be passed directly to the stacks.

Before entering the economizers, the feed water is passed through auxiliary heaters, which receive heat from the exhaust steam from the steam-driven boiler feed pumps and circulating pumps.

Blowers will be used to furnish forced draft under the grates. There will be twenty blowers, each having a capacity of 56,000 cubic feet of air per minute at a pressure equivalent to two inches of water. They are each driven by a fifty-one-H.P. compound steam engine.

Ten steam-driven vertical feed pumps are provided for the present installation. They are located in the engine room adjacent to the division wall between the engine and boiler rooms, one opposite each engine unit.

These pumps are vertical compound duplex with outside packed plunger. The steam cylinders are twelve and seventeen inches in diameter and fifteen-inch stroke. The water cylinders are eight and one-half inches in diameter.

The coal is unloaded from barges by a movable coal tower which runs on a track placed on the pier at the foot of 58th Street. The coal is taken from the barge by a steel clam-shell bucket which hoists the coal to the top of the tower. This bucket is raised by a 200 horse-power direct current 250-volt motor directly connected to the hoisting drum, and is unloaded into a hopper which delivers the coal to crushing rolls, thence to weighing hoppers.

Beneath the track upon which the tower runs is a belt conveyor, upon which the coal is delivered from the weighing hoppers. This conveyor takes the coal to the bulkhead line and unloads upon a similar belt conveyor which runs through a tunnel under 58th Street to the westerly end of the power station, where the coal is elevated by a series of inclined conveyor belts to the level of the top of the coal bunkers.

The coal bunkers have a capacity of 18,000 tons; they are of steel construction with concrete arches and lined with cement. Iron chutes take the coal from the bottom of the bunkers to the firing floor in front of the boilers. Two lines of flight conveyors are also provided and placed beneath the coal bunkers, so that different grades of coal, placed in individual bunkers may be distributed to any or all boilers as desired.

The ashes drop from the grates into hoppers, below which are tracks for steel ash cars drawn by storage battery electric locomotives. The cars are run to the west end of the building and across Twelfth Avenue to a dumping pit, from whence the ashes are carried to an ash bin on the bulkhead line by an ash conveyor. From the ash bin, the ashes are loaded into barges.

All of the coal and ash handling machinery is driven by electric motors.

The engine room is provided with two electric cranes, one of which has two fifty-ton hoists, with one auxiliary hoist of ten tons. The other crane has a twenty-five-ton main hoist and a five-ton auxiliary hoist.

The nine 5,000-kilowatt Westinghouse alternators ordered are designed for delivering three-phase current at 11,000 volts, twenty-five cycles, at seventy-five revolutions a minute. These machines are identical in size, and, with the exception of a few details, the same in construction as those installed in the Manhattan station of the Interborough Rapid Transit Company.

Attention may be called to the frame of the external armature, which has a small keystone section so arranged that, in case it is necessary to remove one of the field spools, this section only need be lifted out by the crane.

The armature windings differ in design from those of the Manhattan machines, the conductors being formed into U-shaped coils, which are slipped through the partially closed slots from both sides of the armature. The ends of the U-shaped conductors are bent and soldered together, forming closed coils, the ends of which are connected to other similar coils.

For excitation of the alternator fields, there are provided five direct current generators, each having an output of 250 kilowatts at 250 volts. Two of these units are directly connected to 400 horse-power marine type engines. Three units are driven by 365 horse-power, 400-volt, induction motors.

A storage battery of a capacity sufficient to carry the exciting load of the station for one hour will also be provided. A motor-driven booster will be furnished for charging this battery.

The auxiliary direct current motors in the plant are driven from one of the exciter generators, and for this purpose each generator is connected through double-throw switches, so that it may be connected either to the exciter or to the auxiliary bus bars.

The switching apparatus is placed on galleries built along the 59th Street wall of the engine room. Diagramatically, the arrangement of oil switches, bus bars and feeder cables is similar to that adopted in the Manhattan station, and relays, operating on overload and reverse current with time limits are provided on corresponding oil switches.

All of the oil switches and group bus bars are located

on the main floor. The main bus bars are arranged vertically in two lines of brick compartments extending the entire length of the power station, placed on a gallery below the oil switches.

The switchboards are placed near the centre of the

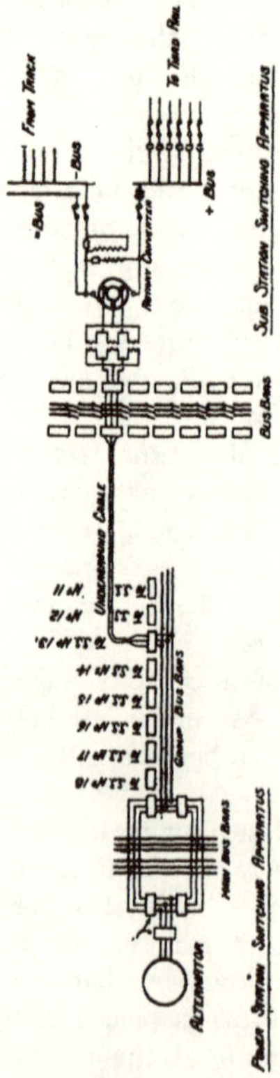

Diagram of System of Power Transmission
Subway Division

present building on a gallery about thirty feet from the engine room floor. Separate instrument and controlling boards are provided for twenty-five-cycle power circuits, and for sixty-cycle lighting circuits.

The arrangement of operating switches, miniature

bus bars and indicating lamps, and other apparatus on the controlling boards, is similar to that in the Manhattan station. The instrument boards are also similar, but have one ammeter only for each feeder cable.

The general illumination of the engine room is provided by Nernst lamps arranged in clusters supported from the roof trusses and single lamps placed around the walls about twenty-five feet from the floor.

In addition, sixteen candle-power incandescent lamps are placed on the engines and along the galleries. The basement is lighted by incandescent lamps. The general illumination of the boiler room is supplied by a row of Nernst lamps arranged one in front of each battery of boilers, and, in addition, there are lines of incandescent lamps in the passage ways around the boilers and incandescent lamps at gauges and water columns.

The boiler room basement, pump room, economizer floor, coal bunkers and conveyor lines are lighted by incandescent lamps. Arc lamps are used around the coal tower and dock.

The lights on the engines and those at the gauge glasses and water columns of the boilers and on the pumps are supplied from the direct current exciter circuits. The other incandescent lamps and the Nernst lamps are supplied with current from three seventy-five kilowatt transformers connected to the sixty-cycle lighting system.

UNDERGROUND HIGH TENSION CABLES

Power is transmitted to the substations through ooo B&S gauge three-conductor cables insulated with $\frac{7}{32}$ of an inch paper around each cable and $\frac{7}{32}$ o an inch paper around the group with $\frac{9}{64}$ of an inch lead sheath. These cables are placed in vitrified clay ducts built in the side walls of the Subway.

Suitable manholes are provided, which open into the Subway and also into the street. In these manholes the cables will be covered with wrappings of asbestos in order to protect adjacent cables in case of a burn-out.

The number of feeder cables running to each substation is the same as the number of 1,500 kilowatt rotary converters installed in the substation. The specified tests on these cables are the same as those for the Manhattan Division.

The longest distance current is transmitted over these cables in 8.2 miles, and the total number of miles of cable is 147.

SUBSTATIONS

There are eight substations, located four on the main line and two on each branch at an average distance of about 12,000 feet apart.

These substations are designed to occupy an area 50 by 100 feet, but individual stations vary somewhat from these dimensions. Foundations are provided for eight 1,500-kilowatt rotary converters in each station.

The buildings are designed with a centre aisle, lighted by sky light and monitor windows. The converters are placed on either side of this aisle with transformers located on the same floor next to the side walls.

The high tension cables are carried under the basement floor to the rear end of the basement, where the end bells and static arrester compartments are located.

The oil switches and bus bar compartments are located on the main floor immediately above, and the alternating current and direct current switchboards are placed on a gallery across the rear of the building.

The steel work of the building is designed with sufficient strength so that two floors may be added for storage batteries, should their installation be found advisable. The front wall of the building and the first bay is already built to the height of these two battery floors.

The rotary converters differ from those installed in the substations of the Manhattan Division in the construction of the frame, which extends down below the floor level. The 550-kilowatt transformers are also nearly identical with those in Manhattan substations, but have less inductance. External inductance is provided which will be connected in the secondary circuit of the transformers should it be desired to operate the converters as compound machines.

The controlling and instrument boards are placed along the front of the switchboard gallery, so arranged that the operator may look down upon the main floor over the controlling bench. The alternating and direct current instruments are placed on a narrow panel supported upon columns about three feet above the controlling board. The alternating current instruments and controlling apparatus are substantially the same as those in Manhattan substations.

The direct current circuit breakers are placed at the rear of the gallery in brick compartments, and are provided with tripping coils operated from the direct current controlling board. The direct current positive bus bar is supported on the back wall of the circuit breaker compartment.

The direct current switches and instruments for the rotary converters and feeders are placed on the operating panel at the front of the gallery. The switches are double throw and provision is made so that, in case the circuit breaker for any feeder or converter is damaged, the current may be transferred to a spare circuit breaker.

Each direct current feeder cable is provided with a switch both at the substation end and at its point of attachment to the contact rail, so that, in case of trouble, the current may be readily cut off from the damaged cable.

The 11,000-volt cables from the oil switches to the transformers are paper insulated, with lead sheath, and are supported on brick shelves built out from the side wall of the basement.

Air for cooling the transformers is furnished by four ninety-inch motor-driven steel plate blowers, driven by 600-volt direct current motors. The passage way between the converter foundations and the side walls is bricked in and serves as an air duct.

CONTACT RAIL, DIRECT CURRENT FEEDER SYSTEM, AND NEGATIVE RETURN

The contact rail weighs seventy-five pounds per yard, and is rolled in sixty-foot lengths. The section is 4⅝ inches high and 4⅝ inches wide at the base. In order to increase the conductivity of the rail, carbon and manganese were made as low as possible. The average composition of the manufactured rails, from mill analyses, was as follows:

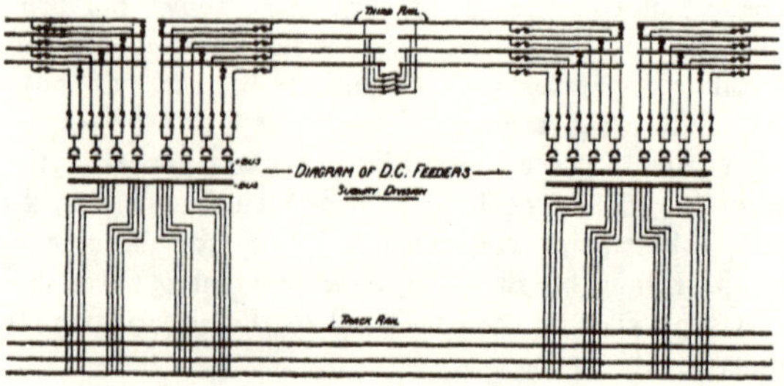

Carbon	.11	per cent.
Manganese	.57	"
Silicon	.02	"
Sulphur	.04	"
Phosphorus	.09	"

The resistance of the rail is about eight times that of an equal section of copper.

The contact rail is supported every nine feet on insulating blocks of "reconstructed granite," which are cemented to a malleable iron pedestal bolted to the ties. Malleable iron clips are clamped to the top of the insulating block and serve to support the rail.

The contact rail centre is located twenty-six inches from the gauge line, and the top of the rail is four inches above the top of track rail.

Each joint is bonded with four 300,000 circular mils stranded copper bonds, with drop forged terminals. Two of these bonds are placed under the splice bars and riveted to the web, and two are riveted to the base of the rail. Each sixty-foot rail is anchored at the centre and the flexible bonds allow for one-half inch expansion at each joint.

The contact rail for each track is divided into two sections between adjacent substations, one of these sections being supplied with current from each substation. At a point half way between substations, the two sections are normally closed through quick-break switches.

The contact rail is supplied with current through feeder cables which are connected to the rail at points along the line, as indicated on the accompanying diagram. This arrangement of the direct current feeder system makes it possible to localize a short circuit on any one of the tracks, and, by opening the switches placed midway between substations, trouble is localized on one track half the distance between substations. The feeder cables are all 2,000,000 circular mils stranded copper cable, insulated with paper and covered by a lead sheath.

For the return circuit, one rail of each track is used, supplemented by copper cables. The other track rail is used for signalling purposes. The negative cables are identical in size and insulation with the positive cables. The track rail is a high carbon rail weighing 100 pounds per yard of American Society standard section in thirty-three-foot lengths.

The joints in one rail of each track, used for the return current, are bonded with two 400,000 circular mils copper bonds, placed under the splice bars.

CAR EQUIPMENT

The five-car trains are to be used for Subway local service, and are composed of three motor cars and two trailers. The eight-car express trains are to have five motor cars and three trailers.

For the operation of the present Subway lines, 700 cars are ordered, of which 430 are motor cars and 270 trailer cars. Unusual precautions have been taken in the design and construction of these cars to make them safe against the risk of fire or mechanical injury. The motor and trailer cars are alike in size and general appearance; they are fifty-one feet two inches long over buffer plates, and twelve feet high above the top of the rail. They have a seating capacity of fifty-two passengers each, with cross and side seats arranged the same as in the Manhattan elevated cars.

The platforms of all cars are vestibuled. At the sides of the platforms, sliding doors are provided, which are operated by levers placed on either side of the front doorway of the vestibule. The doorway in the end of the vestibule of the motor cars is provided with a swinging door, which normally covers the master controller and engineer's valve. The vestibuled platform is used

Subway Motor Car

as a cab on the motor cars, and, when the cab is in use, the swinging door closes the doorway in the end of the vestibule.

Each car is provided with twenty-six 10-candle-power incandescent lamps inside the car body, and two lamps in each vestibule. The motor cars have two destination signals placed at each end of the car roof. Two oil lanterns are placed at each end of the train, the forward lanterns showing white lights, and the rear ones red.

Each car is provided with twenty-four electric heaters, placed in panels under the seats. A heater is also placed in the front of each vestibule in the motor cars. The heater circuits are so arranged that eight, sixteen, or twenty-four amperes may be used in heating the cars.

All of the wiring for heater and lighting circuits is placed in flexible metal conduits. The switches and fuses for heater and lighting circuits are placed on a slate panel mounted in a metal box, which opens into the vestibule at one end of each car. In the motor cars, this panel is made to accommodate also a main switch for motor circuits, control cut-out switch and fuses and air compressor switch and fuses.

The first order was for 500 car bodies, which are identical for motor and trailer cars. These cars have four steel sills extending from end to end of the platforms. The entire undersurface of the car is sheathed with specially treated asbestos board one-fourth of an inch in thickness, and the sides of the car below the windows are sheathed outside with copper in order to eliminate the risk of fire.

An order has recently been placed for 200 additional motor car bodies, which are to be built entirely of steel. These cars are similar in appearance to those above described, and have the same general dimensions. The truss rods are omitted from the steel sub-frame. These cars are finished inside in aluminum and have a singularly light and handsome appearance.

The weight of both types of motor cars is approximately the same, being 73,650 pounds complete, with two motors and without passengers. The weight of trailer car complete without passengers is 51,650 pounds.

The motor trucks are designed to accommodate two 200-H.P. motors each. The wheel base is six feet eight inches, and the wheels are 33.25 inches in diameter.

The trucks conform, in general design, to the master car builder's standard, with swing bolster of cast steel, channel iron transoms with supports for the motor frame, forged steel frame and equalizer bars.

The wheels have steel tires with cast steel centres. The axles are double hammered, open hearth steel, 6.5 inches in diameter at the motor bearings, and $7\frac{13}{16}$ inches in diameter at the gear.

The weight of the truck complete, with gear, but without motors, is 12,500 pounds.

The design of the trailer trucks is similar to that of the motor truck, with five-foot six-inch wheel base and thirty-inch wheels. The axles are four and three-fourths inches in diameter. The weight of the truck complete is 8,900 pounds.

MOTORS

Eight hundred and eighty motors have been ordered for the car equipment. All of these motors are nominally 200 horse-power each, and were specially designed for the Subway service.

The Westinghouse motor, known as No. 86, has a split cast steel frame, the lower half of which is supported at the front end by a bracket on the truck transom. The motor has four equal field coils which are wound with copper ribbon on edge and insulated with mica and asbestos and placed in a water-tight steel shell.

The armature windings are insulated with mica and are capable of withstanding great heat. The weight of the motor complete, with gear and gear case, is 5,900 pounds.

The General Electric No. 69 motors have a cast steel frame in one piece, and the general design is similar to the General Electric No. 66 motors used on the cars of the Manhattan Division. The weight of the motor complete, with gear and gear case, is 5,750 pounds.

The Sprague-General Electric Company's multiple unit control apparatus, used on all the motor cars, is substantially the same as that adopted for the cars on the Manhattan Division. An additional device has, however, been placed in the master controller which regulates the rate of acceleration. A relay is connected in the circuit of one of the two motors on each motor car which prevents the master controller from turning on to successive resistance combinations until the current in every relay in the train has dropped to the predetermined amount. The relays are normally set at 310 amperes.

The master controller is provided with attachments so arranged that if the motorman removes his hand from the handle the current is at once shut off and the brakes are applied. To accomplish this, a button is placed on the controller handle, which must be pressed down by the motorman. If the motorman removes his hand and releases the button, the current is cut off the motors and the air brakes are applied.

A circuit breaker is provided in the main circuit on each motor car, furnished with a tripping coil which operates on overload and may also be operated by the motorman. It has also a resetting coil arranged so that the motorman may close the circuit breaker. The motor circuit is also provided with a 750-ampere copper ribbon fuse, placed under the car.

Stranded cables, insulated with rubber thirty per cent. pure Para and covered with asbestos braid, are used on all the motor circuits. On the motor cars first equipped, this cable is placed in asbestos moulding one-fourth of an inch thick. On the steel cars, the cable is drawn into iron pipes.

The cables are protected by enclosed fuses placed immediately above each contact shoe. A bus line cable, protected by copper ribbon fuses, is connected throughout the train to all the contact shoes.

The wiring and cables on all cars are subjected to a test of 2,000 volts, alternating, after installation.

The cars are equipped with Westinghouse automatic air brakes. Motor-driven air compressors supply air for the operation of the brakes.

Automatic governors are arranged so that all compressors on the train are started whenever any governor closes the circuit.

SUBWAY LIGHTING SYSTEM

Current for lighting the Subway passenger stations and the tunnel itself is supplied from the three 1,250-kilowatt turbo-generators of the Westinghouse-Parsons type at the 58th Street power-house. These generators deliver three-phase current at 11,000 volts potential, and a frequency of sixty cycles. The entire generating and distributing lighting system is independent of the power equipment.

For the distribution of the current from the 58th Street power station, three No. 6 B&S gauge three-conductor cables are employed.

At each underground passenger station, a transformer is installed in a brick compartment built for the purpose. These transformers have two secondary windings—one of 600 volts and the other either 115 or 230 volts. The 600-volt winding supplies sixteen candle-power lamps, wired five in series, attached to the columns in two rows at intervals of about sixty feet along the tunnel; also a few lamps on the stairways, platforms and ticket offices of each station. In case of the burnout of a lighting transformer, the break-down of a high tension lighting cable, or any part of the generating system, these 600-volt lighting circuits may be supplied with 600-volt direct current from the third rail by a double-throw switch.

The primary winding is for 11,000 volts, with two loops brought out so that an adjustment of about five per cent. range can be made to compensate for varying distances from the power station.

In the transformer compartment at each passenger station, between the end bell of the 11,000-volt cable and the transformer, are installed three single-phase, single-throw disconnecting switches, and four-ampere, 11,000-volt, enclosed fuses for the protection of the transformer.

The total number of lights on the entire system is about 8,700, of which about 3,000 are for the illumination of the tunnel. Each underground passenger station is provided with about 120 lights, of which about fifty are thirty-two candle-power incandescent lamps for illuminating the platforms, and the remaining seventy are sixteen candle-power lamps for illuminating the ticket offices, stairways, toilets, etc.

All wiring is done in loricated conduit. The conduit work was installed before the completion of the interior station decorations, and is imbedded in the side walls and ceilings.

The neutral points of all transformer secondary windings are permanently grounded, and a grounded metallic shield is interposed between the 11,000-volt primary and the secondary windings to avoid any possible passage of the high tension current into the secondary wiring.

Panel heaters are provided in all underground passenger stations to warm the ticket offices and closets. Direct current for these heaters is supplied from the contact rail at 600-volts, and controlled from the same panel board which controls the lighting circuits.

AUTOMATIC BLOCK SIGNAL AND INTERLOCKING SYSTEM

The Subway lines are being equipped with an electro-pneumatic block signal system, supplemented by automatic devices which are arranged to set the emergency brakes on any train which may disregard a danger signal. The system includes, also, electro-pneumatic devices for the operation and interlocking of track switches and switch signals.

An alternating current track circuit is used for operating the signal relays, in order that they may be irresponsive to the differences in direct current potential occurring in the track rail. The power for energizing the signal rail and for lighting the signals is supplied by transformers placed at the exit end of each block. The transformer primaries are connected across 500-volt, sixty-cycle mains, which are supplied with power from seven 30-kilowatt motor-generator sets located in the substations.

The signal relays above mentioned serve to open or close local circuits controlling the valves of the air cylinders which operate the signals. These local circuits and the circuits for operating the admission valves for switches and automatic safety stops are supplied with current at sixteen volts from storage batteries.

A two-inch air pipe runs throughout the length of the line for supplying air at sixty to seventy-five pounds pressure for the operation of the pneumatic cylinders in signals, switches and automatic stop devices. Air is supplied to this pipe by eight thirty-five horse-power motor-driven air compressors located in the sub-stations.

APPENDIX

Appendix

THE AMERICAN INSTITUTE OF ELECTRICAL ENGINEERS

In the spring of 1884 the founders of the American Institute of Electrical Engineers conceived the idea of establishing a permanent electrical society. A circular letter was mailed to the electrical engineers of the United States and Canada, inviting their co-operation. The first three paragraphs of this letter were as follows:

"The rapidly growing art of producing and utilizing electricity has no assistance from any American national scientific society. There is no legitimate excuse for this implied absence of scientific interest, except it be the short-sighted plea that every one is too busy to give time to scientific, practical and social intercourse, which, in other professions, has been found so conducive to advancement.

"The American Societies of Civil, Mechanical, and Mining Engineers which have been so prosperous and of such great advantage to their members, are good examples.

"An International Electrical Exhibition is to be held in Philadelphia next autumn, to which many of the famous foreign electrical savants, engineers, and manufacturers will be visitors; and it would be a lasting disgrace to American electricians if no American national electrical society was in existence to receive them with the honors due from their co-laborers of the United States."

This appeal was signed by 93 of the men prominent in the electrical field at that time, most of them connected with the various telegraph systems. About 150 men responded, and a preliminary meeting was held on April 15th, 1884, at the headquarters of the American Society of Civil Engineers. On May 12th, 1884, a second meeting was held and a permanent organization effected.

The first meeting of the Institute for the reading and discussion of professional papers, was held at the Continental Hotel, Philadelphia, October 7-8, 1884. Annual meetings were subsequently held at New York in 1885 and 1886. On June 8th, 1886, the first regular monthly meeting was held at the Mills Building, 15 Broad street, New York. In the autumn of 1886, the regular monthly meetings were resumed and have been continued regularly, excepting July and August each year, since that time. In addition to the regular monthly meetings an annual convention is held each year some time between May and December; the last convention was held at Niagara Falls, June 29 to July 3d, 1903. The next conventionn will be held at St. Louis, Mo., on September 14th, 1904.

All the papers read at Institute meetings are subject to the approval of both the Papers and Editing committees. This policy is rigidly adhered to. The papers are invariably printed and distributed in advance of the meeting at which they are to be read and are printed in the Proceedings which are issued monthly. At the end of the calendar year, these are bound in cloth and distributed without cost to the entire membership.

The officers of the Institute consist of a President, six Vice Presidents, twelve Managers, a Secretary and a Treasurer.

The affairs of the Institute are managed by a Board of Directors, consisting of the officers of the Institute and the two junior Past-presidents.

There are three classes of members, known as honorary, full and associate members. A full member must have been an associate and must be not less than 27 years of age.

The twentieth anniversary of the founding of the

Institute occurred on May 12th of this year. On May 12th, 1884, there were 97 charter members; on July 19th, 1904, there were 3,301 members and associates.

At its September meeting in 1902, the Board of Directors appointed a Committee on Local Organizations for promoting the organization of local meetings. The primary purpose of these meetings was stated to be the presentation of the papers and discussions of the regular meetings of the Institute and for the discussions of them, to be supplemented, however, by new papers, especially on subjects of local interest.

All the important work of the Institute is done by committees appointed by the President. There are six standing committees; these committees are in close touch with the Secretary, and with him outline the policy and direct the routine work of the Institute.

The special committees vary from time to time in personnel and number. At present there are 21 of these committees. Each committee is appointed for a special purpose; it reports to the President and Board of Directors through the Secretary.

In the winter of 1889-90, the American Society of Mechanical Engineers engaged in an effort to provide itself with suitable quarters, and the proposition was made to the Institute that should a house be obtained, the headquarters of the Institute should be established in the same building. Under these conditions the mechanical engineers bought, in June, 1890, the house of the Academy of Medicine at 12 West 31st street. Institute headquarters were established there until 1894, when, needing more space than could be spared at the Mechanical Engineer's home, the Institute moved to 26 Cortlandt street. Outgrowing these quarters, a move was made to 95 Liberty street in 1901, the seventh floor being required to house the Executive Offices and Library.

On the evening of February 9th, 1903, Mr. Carnegie was the guest of honor at the annual dinner of the Institute, and his interest was aroused in plans which were being discussed for providing the Institute with a permanent home of its own.

On February 14th, he wrote a letter to the four professional engineering societies and the Engineers' Club, stating his willingness to contribute one million dollars for the erection of a union engineering building. This offer was accepted by the American Society of Mining Engineers, American Society of Mechanical Engineers, American Institute of Electrical Engineers, and the Engineers' Club. On March 19th, 1904, Mr. Carnegie increased his original offer by five hundred thousand dollars. Two buildings will be erected, one known as the Union Engineering Building and the other as the Engineers' Club. Plans for these buildings have been accepted and it is expected that both will be finished and ready for occupancy by May 1st, 1906.

COLUMBIA UNIVERSITY

Columbia University is the principal seat of higher education in New York city. It is splendidly situated on Morningside Heights, at Amsterdam avenue and 116th street.

At the present time about 5,000 students are enrolled and its teaching force numbers 485.

In the Scientific Department, the Columbia School of Mines has long been recognized as among the leading schools of the world in this branch of professional training. The Electrical Department, organized more recently, has a large teaching personnel and very complete laboratory facilities.

EDISON'S LABORATORY

The experimental laboratory of Mr. Thos. A. Edison is situated at Orange, New Jersey, and may be reached by the Delaware, Lackawanna & Western R. R. to Orange Station, and thence by trolley.

Here Mr. Edison has his technical library and his

chemical and physical laboratories, and it is here that all the experimental work has been done since the laboratory was removed from Menlo Park, about 1884.

THE ELECTRICAL TESTING LABORATORY

Located at 80th street and East End avenue, New York city, the Technical Testing Laboratory contains the most extensive commercial electrical testing equipment in the world.

The life-test for incandescent lamps embraces 6,000 sockets and a motor dynamo outfit to operate the lamps continuously. The photometric laboratories are equipped with all of the most recent devices.

The following manufacturers in New York City and its vicinity have expressed their willingness to allow visitors to see their works:

THE A. B. SEE ELECTRIC ELEVATOR COMPANY

The A. B. See Company makes a specialty of electric elevators. Its new plant, which will be in full working order by September 1st, is located in Jersey City. It is best reached by ferry to Communipaw.

COOPER-HEWITT ELECTRIC COMPANY

The Cooper-Hewitt Company manufactures the Hewitt Mercury Vapor Lamp and Vapor Converter.

The works are located at 220 West 29th street.

CROCKER-WHEELER COMPANY

The Crocker-Wheeler Electric Company is the American licensee of Brown, Boveri & Cie, manufacturers of electrical generators and motors for every variety of service.

The works are located at Ampere, New Jersey, on the Delaware, Lackawanna & Western R. R.

FOOTE, PIERSON & CO.

Successors to E. S. Greeley & Co., manufacturers of telegraph instruments, and supplies, also high grade measuring and testing instruments, X-ray apparatus and electrical specialties.

Office and works at 82 Fulton street, New York city.

MARINE ENGINE AND MACHINE COMPANY

The Marine Engine and Machine Company manufactures electric elevators, controllers, refrigerating machinery, alco-vapor launches, heavy machine castings, etc.

The plant is located in Harrison, New Jersey, near Newark. It may be reached by the Pennsylvania Railroad or the Delaware, Lackawanna & Western R. R.

OTIS ELEVATOR COMPANY

The Otis Elevator Company manufactures hydraulic, electric and steam elevators, also hoisting engines, escalators and inclined railways.

The works of this company are located in Yonkers.

WARD LEONARD ELECTRIC COMPANY

The Ward Leonard Electric Company manufactures rheostats, motor starters, circuit breakers, resistance units, etc., the manufacture of various forms of rheostats being its speciality.

The works are located about 13 miles from the Grand Central station, on the Harlem Branch of the New York Central & Hudson River R. R.

WESTERN ELECTRIC COMPANY

The Western Electric Company manufactures telephonic apparatus of all kinds, including transmitters, receivers, switchboard apparatus, etc.

The New York factory is located at the corner of West and Bethune streets, New York city.

THE WILLYOUNG & GIBSON COMPANY

The Willyoung & Gibson Company manufactures electrical and scientific instruments for testing and other purposes. It also manufactures all kinds of X-ray coils.

The plant is located at 40 West 13th street.

www.ingramcontent.com/pod-product-compliance
Lightning Source LLC
LaVergne TN
LVHW101439200926
848483LV00004B/407

* 9 7 8 1 4 1 8 1 7 1 6 7 4 *